The Truth Machine

JOHNS HOPKINS STUDIES IN THE HISTORY OF TECHNOLOGY
Merritt Roe Smith, Series Editor

The Truth Machine

A Social History of the Lie Detector

Geoffrey C. Bunn

The Johns Hopkins University Press
Baltimore

© 2012 The Johns Hopkins University Press
All rights reserved. Published 2012
Printed in the United States of America on acid-free paper
2 4 6 8 9 7 5 3 1

The Johns Hopkins University Press
2715 North Charles Street
Baltimore, Maryland 21218-4363
www.press.jhu.edu

Library of Congress Cataloging-in-Publication Data
Bunn, G. C. (Geoffrey C.)
The truth machine : a social history of the lie detector / Geoffrey C. Bunn.
p. cm. — (Johns Hopkins studies in the history of technology)
Includes bibliographical references and index.
ISBN-13: 978-1-4214-0530-8 (hdbk. : alk. paper)
ISBN-13: 978-1-4214-0651-0 (electronic)
ISBN-10: 1-4214-0530-x (hdbk. : alk. paper)
ISBN-10: 1-4214-0651-9 (electronic)
1. Lie detectors and detection—History. 2. Lie detectors and detection—United States—History. I. Title.
HV8078.B86 2012
363.25′4—dc23   2011044971

A catalog record for this book is available from the British Library.

Special discounts are available for bulk purchases of this book. For more information, please contact Special Sales at 410-516-6936 or specialsales@press.jhu.edu.

The Johns Hopkins University Press uses environmentally friendly book materials, including recycled text paper that is composed of at least 30 percent post-consumer waste, whenever possible.

What therefore is truth? A mobile army of metaphors, metonymies, anthropomorphisms: in short a sum of human relations which become poetically and rhetorically intensified, metamorphosed, adorned, and after long usage seem to a nation fixed, canonic and binding; truths are illusions of which one has forgotten they *are* illusions, worn-out metaphors which have become powerless to affect the senses; coins which have their obverse effaced and now are no longer of account as coins but merely as metal.

—Friedrich Nietzsche, "On Truth and Lie in an Extra-Moral Sense" (1873)

It is so easy to do wrong! Everything the Devil makes runs easily. It is only God's machinery which has friction. The lie is spontaneous;—the truth requires thought. Yet the offhand production is born with the seeds of decay in it, and its other name is "Death." Its history is always cyclical, and returns upon itself; for the path of a lie is so tortuous that, sooner or later, it is bound to intersect its own course. Then comes discovery, humiliation, pain—retribution. The hyperbola of deception has never yet been plotted.

—Milton L. Severy, *The Mystery of June 13th* (1905)

CONTENTS

Introduction
Plotting the Hyperbola of Deception 1

Chapter 1
"A thieves' quarter, a devil's den":
The Birth of Criminal Man 7

Chapter 2
"A vast plain under a flaming sky":
The Emergence of Criminology 30

Chapter 3
"Supposing that Truth is a woman—what then?":
The Enigma of Female Criminality 51

Chapter 4
"Fearful errors lurk in our nuptial couches":
The Critique of Criminal Anthropology 75

Chapter 5
"To Classify and Analyze Emotional Persons":
The Mistake of the Machines 94

Chapter 6
"Some of the darndest lies you ever heard":
Who Invented the Lie Detector? 116

Chapter 7
"A trick of burlesque employed . . . against dishonesty":
The Quest for Euphoric Security 134

Chapter 8
"A bally hoo side show at the fair":
The Spectacular Power of Expertise   154

Conclusion
The Hazards of the Will to Truth   174

Acknowledgments   193
Notes   195
Essay on Sources   237
Index   241

The Truth Machine

INTRODUCTION

Plotting the Hyperbola of Deception

> An increased liberalism in the definition of "fact" can have grave repercussions, while the idea that truth is concealed and even perverted by the processes that are meant to establish it makes excellent sense. —Paul Feyerabend, *Against Method* (1975)

On January 30, 1995, not long after O.J. Simpson had released *I Want to Tell You*, the book he hoped would clear his name, the tabloid television show *Hard Copy* revealed that they had subjected the double murder suspect to a lie detector test. The former football star had recorded himself on tape, reading aloud various passages from his book: "I want to state unequivocally that I did not commit these horrible crimes."[1] *Hard Copy* hired lie detector expert Ernie Rizzo to use a "Psychological Stress Evaluator" to subject Simpson's voice to stress analysis. According to the show's "Hollywood Reporter," Diane Dimond, the test could separate "fact from fiction." Used by the police, the military, and big business, the instrument had been shown to be "95 percent accurate." As a result of Rizzo's analysis, he concluded that Simpson was "one hundred percent deceitful... one hundred percent lying."[2] One week after *Hard Copy*'s deception test, supermarket tabloid newspaper the *Globe* subjected the same tape recording of Simpson's voice to "Verimetrics," a high-tech lie detector favored by police investigators.[3] But this time Jack Harwood, a "Veteran investigator," proclaimed Simpson "absolutely truthful," noting that the "lie test shows O.J. didn't do it!"

One type of lie detector, identical statements from a single suspect, and two equally emphatic yet contradictory verdicts. When Simpson said, "I would take a bullet for Nicole," Harwood claimed, "the former football hero was being completely honest," while according to Rizzo he was "absolutely lying." How can two experts both claim scientific validity for their respective instruments, analyze the same material, and reach completely different conclusions?

Early histories of the lie detector celebrated the many famous and infamous cases in which it had been used during the twentieth century.[4] More recent studies have either challenged the instrument's scientific status, or questioned its legitimacy on grounds that this practice constitutes an assault on civil liberties.[5] David Lykken was one of the first psychologists to dispute claims about the machine, arguing, "the lie detector has no more place in the courts or in business than a psychic or tarot cards."[6] According to Lykken, by 1980 more than one million lie detector tests were performed annually in the United States.[7]

The classic polygraph examination involves simultaneously measuring a suspect's blood pressure, breathing rate, and electrical skin conductance as a series of questions that require yes or no answers are asked. But the person can also be subjected to more covert scrutiny: "behavior symptoms" are observed before and after the test is performed; cameras behind two-way mirrors may record gestures and nuances of expression. Talkativeness and enthusiasm may be noted, to be incorporated into the examiner's final assessment of truth or deception. It seems that no lie detector examination takes place under "objective" scientific conditions divorced from the wider social context. And symbols lend insight into the values that underscore the lie detector test. What better emblem of masculine professional power than the *briefcase*, that mandatory accessory of every polygrapher? From the black briefcase comes the *chart*, at once a graphic calculus of guilt and a sacred scroll inscribed with the truth. Consider also the *chair*, a seat for the sovereign subject with whom no eye contact must be made, but also a constraining device, reminiscent of the electric chair.

The demarcation between the supposed rationality of the male polygrapher and the supposed apparent emotionality of a female subject is a salient feature of lie detector discourse. The instrument was designed to reveal the supposed invisible pathologies of the female body, an approach with a long precedent in criminology, a history that this book examines. For the science

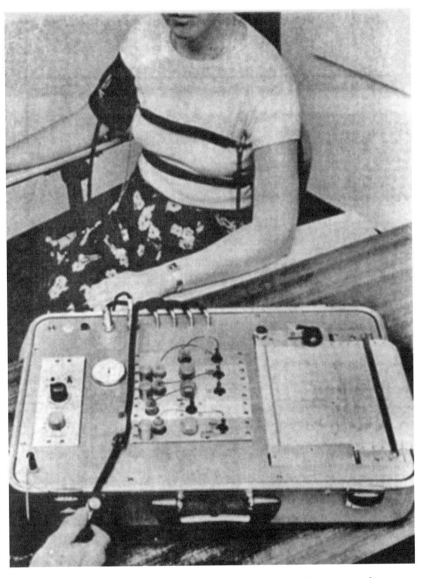

The traditional polygraph measures skin conductance, breathing rate, and blood pressure. The subject also undergoes intense visual scrutiny.

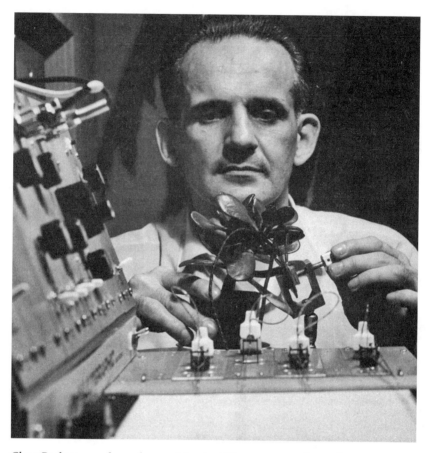

Cleve Backster, a polygraph expert for the CIA, attempts to detect deception in a plant. Photo by Henry Groskinsky, Life.com images.

of "pupillometrics"—the attempt to detect dishonesty by recording changes in pupil size—the gaze of the subject becomes the important characteristic of the deception test. In a recapitulation of criminal anthropology's fruitless search for visible stigmata of criminality, almost every body part has been subjected to testing: the hand, arm, skin, lungs, heart, muscles, voice, stomach, and brain have all been examined at some point in the history of this technology. Sometimes it has not just been the human body that has attracted pioneers. In the late 1960s, Cleve Backster achieved international notoriety for attaching his polygraph to a philodendron plant, claiming it could detect "apprehension, fear, pleasure, and relief."[8] A former Central Intelligence

Agency interrogator and director of the Leonarde Keeler Polygraph Institute of Chicago, it was Backster who introduced the "Backster Zone Comparison Polygraph," which became the standard polygraph model used at the U.S. Army's Polygraph School. By 1969 it seems he had single-handedly created the urban legend that plants had emotions: "We have found this same phenomenon in the amoeba, the paramecium, and other single-cell organisms, in fact, in every kind of cell we have tested: fresh fruits and vegetables, *mold* cultures, yeasts, scrapings from the roof of the mouth of a human, blood samples, even spermatozoa."[9]

Between the 1935 Lindbergh "crime of the century" and the 1995 O.J. Simpson "trial of the century," the notion of the lie detector became deeply embedded in the North American psyche. Despite constant criticism, satirical attacks, government prohibition, Papal condemnation, and a widespread suspicion that it "can be beaten," the use of the lie detector persists. High-profile cases in which the participants took polygraph tests include cases involving Anita Hill and Clarence Thomas, the spy Aldrich Ames, and the Oklahoma City and Atlanta Olympics bombings. Isuzu trucks, Pepsi Cola, and Snapple juice are some of the products that were advertised with the help of the "truth machine." It appeared in countless movies and television shows. In one *Star Trek* episode, Captain Kirk made Scotty take a lie detector test to prove he was not a serial murderer of women. An episode of the 1990s hit TV cop show *Homicide* featured "the electro-magnetic neutron test" which, unknown to the suspect, issued photocopies of a palm print upon which the words "True" or "False" had been printed beforehand. Many such depictions, of course, portray lie detection as a rational and technical science by contrasting it with "pseudoscience." But use of the machine has constantly transgressed the boundary that supposedly demarcates factual science from sheer fantasy.

The use of the lie detector to manage contradiction is a key theme of this book, one that previous histories have not highlighted. The principle ambition here is to investigate how the lie detector came to be constructed as a technology of truth.[10] Why do these machines continue to feature in the dreams of those responsible for maintaining law and order? Recent scholarship has detailed the biographies and motivations of the major actors in the historical drama.[11] My aim is to push the story back in time into the obscure origins of criminology itself. What interests me is why and how the lie detector was finally "invented" in the United States, even though all the important technological innovations had been developed by European criminologists prior to the start of the twentieth century. My argument is that the machine came about as the result of a sustained dialogue between science—in this case criminology—and the wider culture. Literary, newspaper, and movie depictions did not misinterpret, distort, or corrupt the concept of the lie detector; in fact they played a vital role in creating it.

CHAPTER 1

"A thieves' quarter, a devil's den"
The Birth of Criminal Man

> There is a thieves' quarter, a devil's den, for these city Arabs. There is their Alsatia; in the midst of foul air and filthy lairs they associate and propagate a criminal population. They degenerate into a set of demi-civilized savages, who in hordes prey upon society . . . a race as fierce as those who followed Attila . . . These communities of crime, we know, have no respect for the laws of marriage—are regardless of the rules of consanguinity; and, only connecting themselves with those of their own nature and habits, they must beget a depraved and criminal class hereditarily disposed to crime. Their moral disease comes *ab ovo*.
> —J. Bruce Thomson (1870)

For much of the last two millennia in the West, the Christian tradition considered the miscreant's deeds to be manifestations of universal sin. Human weaknesses such as depravity, temptation, lust, and avarice were regularly invoked to account for conduct that compromised the moral order. Criminality was explained by appealing to supernatural forces such as the actions of mischievous demons or the vagaries of fate. People in early modern England believed that God exposed and punished the crime of murder either through direct intervention or by acting through temporal agents.[1] The supreme power of divine providence guaranteed that crimes of blood would be punished, despite the difficulties associated with detection and proof.

In 1591, a Kent coroner ordered the murderers of four children to call out their names, whereupon the victims' pale bodies, "white like unto soaked flesh laid in water, sodainly received their former coulour of bloude, and had such a lively countenance flushing in theyre faces, as if they had beene living creatures lying aslepe, which in deed blushed on the murtherers when they wanted grace to blush and bee ashamed of theyre owne wickednesse."[2] In the 1650s, Lady Purbeck and a maidservant were both instructed to lay their hands on the corpse of an infant discovered in a privy. The maidservant immediately confessed to the murder when the body started bleeding.[3] In 1725, London magistrates ordered that a human head found on the shore of the Thames at Westminster be placed on a pole in a nearby churchyard, directing church officials to arrest anyone "who might discover signs of guilt on the sight of it."[4] Locals recognized the head as that of John Hayes, whose wife, the magistrates quickly discovered, had recently taken two lovers. Convicted of the murder, Catherine Hayes was subsequently burned alive. As these examples demonstrate, in the early modern period, corpse touching, cruentation rituals,[5] and violent public executions appeared to materialize divine intelligence; they produced confessions and acted as powerful deterrents against crime.

Such procedures were products of mental frameworks and ways of life very different from those of our own era.[6] Modern ideas about criminality originated in the Enlightenment of the eighteenth century. Cesare Beccaria in Italy, Jeremy Bentham and John Howard in Britain, Benjamin Rush in America, and Paul Johann Anselm von Feurerbach in Bavaria pursued rational inquiries into the causes of crime and prison reform, pioneering a secular, modernist criminological discourse.[7] Rush's *The Influence of Physical Causes upon the Moral Faculty* (1786) was one of the first scientific attempts to conceptualize crime and insanity in terms other than sin.[8] Bentham described his innovative model prison, "Panopticon" of 1785, as "a new mode of obtaining power of mind over mind, in a quantity hitherto without example." The ambition of what later became known as the "classical school" was that "government by rule" would replace "unregulated discretion," in the words of one magistrate reformer.[9] Such beliefs were based on a rationalist conception of the calculating subject, a model of the individual whose "psychological" motivations were irrelevant to the administration of justice. Under this framework, research into the biological or environmental causes of crime would only undermine the liberal conviction that individuals were autonomous agents in full control of their own actions.[10]

The general thrust of penal policy during the nineteenth century was humanist. The number of hangings declined in England from the 1830s, and by the 1860s all the traditional Georgian penalties had been abandoned, including the pillory and whipping post, the convict ship, and the public execution.[11] Public violence was increasingly thought to pander to the lowest human instincts and militate against improving the conduct of the population. Punitive spectacles were gradually replaced by measured disciplinary techniques. The prison was rejuvenated as a space for moral discipline, "a training ground for, and a social representation of, the overcoming of immediate impulses and passions and the reconstruction of character."[12] Punishment strategies that had once broken the body transmuted into to those that promised to repair the mind. These reforms brought the hitherto indistinct image of the criminal into sharp focus.

At the start of the nineteenth century, the criminal had been little more than "a pale phantom, used to adjust the penalty determined by the judge for the crime."[13] By its end he had eclipsed the crime and had become the focus of criminological discourse. A diverse array of intellectual, scientific, practical, social, and political developments combined to create an empirical discipline devoted to systematically analyzing the causes of criminality.[14] New theories of human nature conceptualized the mind as a natural entity, particularly in the wake of evolutionary theory.[15] The reconceptualization of human agency in the language of naturalism made it possible to think about criminality less in terms of moral failures of the will and more in terms of the mind's constitutional and environmental influences. Scientific explanation began to shift "from acts to contexts, from the conscious human actor to the surrounding circumstances."[16] Geniuses, criminals, and the insane populated the new disciplines—the three types of exceptional people that had inaugurated the anthropological study of human beings in the late eighteenth century.[17] The "insane criminal genius"—that diabolical combination of all three foundational categories of the human sciences—inhabited the pages of learned journals and was thought to stalk the streets of the metropolis.

Another influence on the development of criminology was the emergence of statistics. Parliamentary committees charged with investigating rising crime rates required systematic numerical data about crime. The French government started publishing official crime figures in 1827. In Britain, the Statistical Department of the Board of Trade was founded in 1832, the Manchester Statistical Society the following year. In 1834 the Statistical Society of London began to use judicial statistics and census data to chart the distribution and demography of crime and to correlate crime rates with other social indices.[18] Conclusions about the complex causes of criminal conduct could be framed in a new way.[19] The Belgian statistician Adolphe Quetelet repeatedly cited crime rates as evidence for his claim "that free-will exercises itself within definite limits."[20] "Sad condition of the human species!" he exclaimed in 1835. "We are able to enumerate in advance how many individuals will stain their hands with the blood of their fellows, how many will be forgers, how many prisoners, much like one can enumerate in advance the births and deaths that must take place." Society contained within itself the seeds of all the crimes that were to be committed, he asserted, as well as the conditions necessary for their nurturance: "It is society that, in some way, prepares these crimes, and the criminal is only the instrument that executes them."[21] Quetelet was not proposing a social account of crime. Under the influence of phrenology, the "reluctant determinist" statistician accepted that "unhealthy morality was manifest in biological defects and that those with such defects had high criminal propensities."[22] "I am far from concluding that man can do nothing for his amelioration," he wrote in 1842. "He possesses a moral strength capable of modifying the laws which concern him."[23] Nevertheless, statistics revealed the relative stability and predictability of crime rates and also that age and gender were the two most significant factors determining a person's propensity for criminality.

Public health also contributed to criminology's development. Contemporary observers who noted the unprecedented growth of cities linked pathology with morality by deploying metaphors of disease, sewage, pollution, and contamination. An "avalanche of numbers"[24] enumerated the slums and their inhabitants: the laboring classes, the idle, and the "incorrigibly lazy." In 1856 the great Victorian urban investigator and social critic Henry Mayhew divided society into civilized citizens and nomadic vagabonds. His experiences interviewing beggars, street entertainers, market traders, and prostitutes led him to the "melancholy" conclusion "that there is a large class, so to speak, who belong to a criminal race, living in particular districts of society.... These people have bred, until at last you have persons who come into the world as criminals, and go out as criminals, and they know nothing else."[25] Mayhew claimed that society was composed of two races, "the wanderers and the settlers." Among the former he counted pickpockets, beggars, prostitutes, street performers, sailors, and such like. This group was characterized by "a greater development of the animal than of the intellectual or moral nature of man . . . distinguished for their high cheek-bones and protruding jaws."[26] He considered this group's "habitual indisposition to labour" as the most important cause of crime.

It was a simple step to consider the "dangerous classes" in terms of their deterioration to an earlier stage of biological development, most famously in B. A. Morel's influential *Treatise on the Physical, Intellectual and Moral Degeneration of the Human Race* of 1857. An extended allegory of the Fall, Morel explained how God's original creation could have become corrupted over countless generations. Included in his theory was a Lamarkian account of the origins of stigmata—degenerated physiological and anatomical characteristics: "When under any kind of noxious influences an organism becomes debilitated, its successors will not resemble the healthy, normal type of the species, with capacities for development, but will form a new sub-species, which, like all others, possesses the capacity of transmitting to its offspring, in a continuously increasing degree, its peculiarities, these being morbid deviations from the normal form—gaps in development, malformations and infirmities."[27] Morel's theory facilitated the entry of "moral insanity" into criminological discourse, a term that was soon part of the language of mental disease.[28] Initially coined to label offenders whose behavior appeared mad while their minds remained sane, moral insanity was transformed into a standard, if contested, psychiatric term by James Prichard's *A Treatise on Insanity* (1835).[29] Morally insane defendants should be sent to the asylum, not the gallows, Prichard argued. Thought to affect the emotions and the will rather than the intellect, moral insanity was a catalyst for psychiatry's redefinition of madness as loss of control.

As the nineteenth century progressed, "characterizations of deviance and crime moved from moral to natural categories . . . [appearing] more deeply rooted within the offender's nature than in the moral consciousness or the rational intellect."[30] By the early twentieth century, the explanation of crime in terms of moral weakness had been replaced by scientific, bureaucratic, and literary discourses that privileged naturalistic explanations.[31] Yet the emphasis on naturalistic causality led to an increase, not a decrease, in the opportunities for governance—the targeting of the criminal's body and mind by mechanisms of regulation and control.[32]

An important stimulus for criminology's concept of inherent criminality was phrenology, the art of reading character from the contours of the skull.[33] Emphasizing observation and reasoning about empirical facts rather than divine revelation, it brought about one of the most radical reorientations of ideas about crime and punishment in the Western tradition.[34] As one leading exponent put it in 1836, the science "explains and proves the fact of some individuals being naturally more prone to crime than others."[35] A popular science, phrenology was also a technology of self-fashioning and personal transformation.[36] Although only a few phrenological works were exclusively devoted to crime and punishment,[37] phrenological periodicals regularly featured extensive discussions of case studies of notorious thieves and murderers. Most phrenological practitioners were interested in the criminal. Phrenology's founder, Franz Josef Gall, left a collection of skulls and casts of heads and brains consisting of "103 famous men, 69 criminals, 67 mental patients, 35 pathological cases and 25 exotics (non-European races)."[38] The anonymous author of *The Philosophy of Phrenology Simplified* (1838) discussed the skulls of two infamous villains: one of the grave robber and murderer William Hare (of "Burke and Hare" fame), whose "acts were such as to fill every well constituted mind with horror and disgust"; the other of Pope Alexander VI, a man whose "life was a series of crimes" and whose character was "grossly bestial, without a redeeming amiable quality."[39] Of two women in confinement, the author explained, one imprisoned for stealing, the other for concealing the stolen articles, "the former will have the organ of Acquisitiveness larger . . . while the second will have the organ of Secretiveness much developed." The "chief of a robber band" would have enlarged organs of Self-Esteem and Determinateness whereas the "habitual vagabond thief" could be distinguished from "a coiner of false money by his having, besides the organ of Acquisitiveness, the organ of Locality larger, and smaller organs of Cautiousness and Constructiveness."[40] Thomas Stone, President of the Royal Medical Society, presented data to suggest that Burke's organ of Benevolence was "both *absolutely* and *relatively* above the average size of the same organ" found among a group of thirty-seven law-abiding citizens. Stone considered the claim made by "some of the most distinguished of the Edinburgh Phrenologists . . . that . . . Burke was really a benevolent man" patently absurd: "to argue the point seriously would be to indulge in one of the severest satires that can be conceived, on the incongruity of the phrenological doctrines."[41]

In principle many of the phrenological faculties or "organs," either enlarged or diminished, were believed to predispose an individual toward criminal acts. A robber of churches, for example, would likely be thought to have a smaller organ of Veneration compared to a humble purse-snatcher. Veneration was considered one of two mental organs that were damaged when an iron rod pierced Phineas Gage's brain in a dreadful railway accident in 1848.[42] The injured laborer's subsequent descent into crime was explained in phrenological terms. Quiet and respectful before the injury, after he recovered Gage became "gross, profane, coarse, and vulgar. . . . The iron rod passed through the regions of BENEVOLENCE and VENERATION, which left these organs without influence in his character, hence his profanity, and want of respect and kindness; giving the animal propensities absolute control in his character."[43]

Gall proposed that the faculties of greed, self-defense, and the "carnivorous instinct" could lead to theft, aggression, and murder respectively.[44] One organ was thought to be especially relevant to the discussion of criminality: Destructiveness, the faculty that Gall called "the organ of Murder."[45] One of the "affective propensities," a developed organ of Destructiveness could apparently be found among enthusiasts of hunting and shooting, and "those fond of attending executions, cock-fighting, and such amusements as lead to the severe punishment or probable death of animals."[46] George Combe, whose *Constitution of Man* (1828) did much to popularize the science, suggested that the faculty produced "the impulse, attended with desire to destroy in general."[47] Conspicuous "in the heads of cool and deliberate murderers, and in persons delighting in cruelty," it was also evidently enlarged among satirists, especially those authors "who write cuttingly, with a view to lacerate the feelings of their opponents."[48] Phrenology's flexible explanatory system was part of its great appeal.

The work of the Liverpool phrenologist Frederick Bridges illustrates many of the science's assumptions and values—not to mention its popularity. In his *Criminals, Crimes, and Their Governing Laws, as Demonstrated by the Sciences of Physiology and Mental Geometry* (1860), Bridges protested that the topic had "hitherto, been treated by the fallacious methods of scholastic metaphysics." This error "rendered it impossible to deduce any sound practical system of treatment of this greatly mistaken class of unfortunate beings."[49] Basing his ideas instead "upon the order of nature," Bridges argued that a symmetrical balancing of propensities was necessary for mental harmony and "moral self-government." Were a person's animal propensities and instincts to preponderate over their moral and intellectual faculties, violence and criminality would ensue. "A very large head," wrote Bridges, "where the organism is badly proportioned is a sure sign of a weak character."[50] Bridges invented a "Phreno-physiometer" to assess the extent of any disequilibrium, an instrument that measured the angle subtended by a line drawn from the opening of the ear to the eyebrow and the horizontal. The angle, which he called the "basilar phreno-metrical," allowed Bridges to infer the class a criminal belonged to, "whether it be that of the murderer, the freebooter, the petty thief, the swindler, or the mental and moral class." Bridges hoped his scheme would be of great practical importance in education and the treatment of criminals.[51]

Bridges claimed to have discovered a natural system for classifying criminals. In murderers the angle was thought to be around 40°, in the law-abiding it was a mere 25°. Too small an angle would produce a "tame and useless" person. Bridges predicted that a parcel boy—whose basilar phreno-metrical he

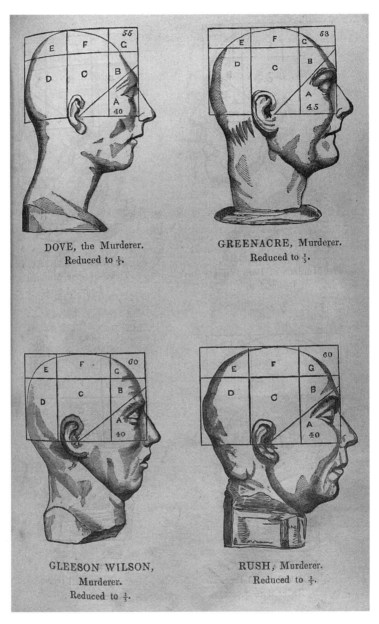

The "basilar phreno-metrical angle" allowed Frederick Bridges to determine which criminal class a person belonged to, "whether it be that of the murderer, the freebooter, the petty thief, the swindler." Frederick Bridges, *Criminals, Crimes, and Their Governing Laws, as Demonstrated by the Sciences of Physiology and Mental Geometry* (London: George, Philip and Son, 1860).

measured to be 38°—would go on to "commit some most diabolical outrage." The boy later set fire to a small child. He also recounted the execution of a certain "Dove of York," who had been convicted of poisoning his wife. Having measured the executed man's basilar phreno-metrical to be 40°, Bridges concluded, "the reflective faculties and moral feelings of the culprit were so small that he was rendered idiotic." "The type of his head is that of a low, vicious, partially mental and moral idiot, who ought not to have been allowed personal liberty."[52] The configuration of his brain, furthermore, did "not range much higher than that of the black monkey."[53] The result was that in Bridges' opinion the execution had been an illegitimate "legal murder."

According to George Combe, because a single mental faculty could become diseased "moral patients... should not be punished, but restrained, and employed in useful labour during life, with as much liberty as they can enjoy without abusing it."[54] Phrenologists also argued against debilitating punishments such as the whip, treadmill, and solitary confinement. They favored prison reform and opposed the death penalty on the grounds that it brutalized onlookers. Some also lobbied for an end to transportation to the colonies, a measure they regarded as devoid of reformative value.[55] The origin of phrenology's reformist agenda lay in its conviction that the human mind had a certain inherent plasticity and could be improved with the appropriate techniques.[56] The notion that the faculties could be changed accounts for the science's enormous appeal among artisans and the aspirational classes.[57] This ethic provided middle-class reformers "with exactly the science they needed to fight their jurisprudential and penological crusades."[58] Phrenologists were active in a wide range of reformist projects such as the slavery abolition movement, temperance movement, and public health campaigns. The science, thus, played an important role in the reform of criminal jurisprudence, opposing retribution and deterrence in favor of reformation.[59]

In Britain, phrenology was but one project among many—such as vegetarianism and sexual purity—that sought to assemble character in alignment with the values of duty, citizenship, integrity, and, above all, self help.[60] Character was not fixed by nature, but was considered to be malleable and susceptible to moral training. For early Victorians, the ambition to build character permeated every field of understanding of human nature, society, and public policymaking.[61] Because it resulted from defective self-management, crime could be treated by developing the offender's psychological capabilities. By the 1860s, the law was being used with increasing consistency as an instrument for developing character and self discipline. The effect of this "civilizing process" was to center the criminal justice system on the ideal of the self-restrained responsible individual.[62]

In his 1869 defense of phrenology, the Coventry philanthropist and "phreno-socialist" Charles Bray reported that the science had influenced at least one senior police officer: "Our most respectable and highly intelligent superintendent of police I found had long been a phrenologist without knowing it. In choosing his men he said he rejected small heads, and chose overhanging foreheads and high heads, as far removed as possible from the criminal type, with which he seemed to be perfectly familiar."[63] Bray approved of this method of selection for the law-abiding but lamented its neglect elsewhere. He reported that as early as 1836 Sir G. S. Mackenzie had unsuccessfully advocated the use of phrenology in the classification of criminals. Mackenzie had petitioned Lord Glenelg, then Secretary to the Colonies: "At present," [Mackenzie] said, "they are shipped off, and distributed to the settlers, without the least regard to their character or history." "There ought," he said, "to be an officer qualified to investigate the history of convicts, and to select them on phrenological principles. That such principles are the only secure grounds on which the treatment of convicts can be founded; proof may be demanded, and it is ready for production."[64] "Of course," Bray complained, "the prayer of Sir George Mackenzie's petition could not be granted. What would all the parsons have said to the doctrine, that the 'differences in moral character are now ascertained to be the effects of difference in organisation!' "[65]

Usually associated with liberal politics, in the hands of Charles Bray and Joshua Toulmin Smith, phrenology could also be used to support radical causes.[66] But a naturalistic approach to human character could be as easily deployed to serve reactionary ends. To a considerable degree criminology was based on the idea that because the criminals' mind exhibited inherent constitutional flaws it was necessary to adopt a vigilant attitude toward their governance. At the 1869 meeting of the British Association, Dr. G. Wilson read a paper, "The Moral Imbecility of Habitual Criminals as Exemplified by Cranial Measurement." Having found that the average size of the heads of 464 criminals was less than that of the ordinary population, Wilson concluded "cranial deficiency is associated with real physical deterioration."[67] A few years later, in the Pavilion of Anthropological Sciences at the Universal Exhibition, a French Professor of Medical Geography claimed to have located a specific type of human being in his collection of thirty-six murderers' skulls. As Girard de Rialle explained to readers of *The Journal of the Anthropological Institute of Great Britain and Ireland*, M. Bordier "was struck by the peculiar formation of these skulls, which all showed characteristics of atavism, and reminded him of prehistoric types. His examination led him to the conclusion that the criminal man is an anachronism, a savage in civilized country, and he compares him to those restive animals which eventually appear in our tame species."[68] Of the thirty-six skulls, only three were neither abnormal nor pathological. Bordier disagreed with British psychiatrist Henry Maudsley that the criminals represented "intermediate types between men sane and insane." He argued that had they been submitted "to a right cerebral orthopedy they would not have been guilty of such crimes."[69]

In 1870, Dr. J. Bruce Thomson, Resident Surgeon of the General Prison for Scotland at Perth, concluded that criminals formed "a variety of the human family quite distinct from civil and social men." Personal experience of criminals over many years had convinced him that "in by far the greatest proportion of offences *Crime is Hereditary*."[70] Thomson noted that having visited the great prisons of England, Ireland, and Scotland, "the authorities, governors, chaplains, surgeons, warders, concur in stating that prisoners, as a class, are of mean and defective intellect, generally stupid, and many of them weak minded and imbecile."[71] The criminal class was thought to possess "a low type of physique, indicating a deteriorated character which gives a family likeness to them all." Thomson agreed with "an accomplished writer" that those born into crime were "as distinctly marked off from the honest industrial operative as 'black-faced sheep are from the Cheviot breed.'" Crime was nothing less than "a moral disease of a chronic and congenital nature, intractable in the extreme, because transmitted from generation to generation." Thomson quoted a Hebrew proverb that had a Lamarckian echo: "The fathers have eaten sour grapes, and the teeth of the children are set on edge."[72]

Fishermen, according to Thomson, thanks to their immobile habits, "intermarry among themselves, and preserve distinct physical and mental characteristics unchanged for centuries." Miners, "who from generation to generation pursue the same calling—form a colony by themselves, and, being the latest of all the industrial classes to emerge from serfdom, are quite a marked variety of men and women."[73] Whereas the "common thief, or robber, or garrotter" possessed "a set of coarse, angular, clumsy, stupid set of features and dirty complexion," clerks, railway officials, and "decent" industrial operatives could be distinguished by their "better physical appearance."[74] Thomson's argument rested on the assumption that criminality ran in families and suggested that the "evil propensities" of one family of five criminals seemed "to have been inherited from the mother; the mother also being a poor silly crea-

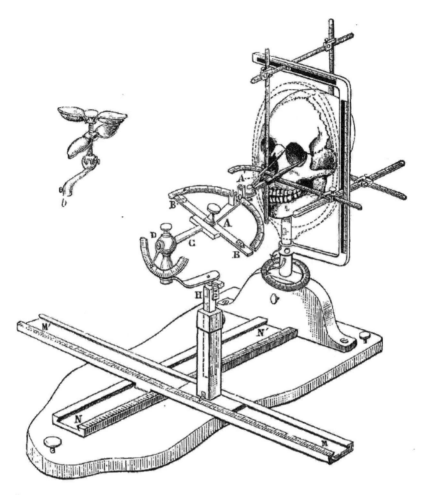

"A savage in a civilized country." Carlo Gaudenzi's instrument for measuring the contours of the skull (1892). Cesare Lombroso, *Les applications de l'anthropologie criminelle* (Paris: Félix Alcan, 1892).

ture."[75] Thomson concluded that crime was "so nearly allied to insanity as to be chiefly a psychological study."[76]

The prison surgeon's zealous diatribe was a link between Mayhew's vivid descriptors of a criminal underclass and Cesare Lombroso's biological typology of the born criminal. While Lombroso was undertaking his celebrated study of Giuseppe Villella's anomalous skull—the event that produced "the totem, the fetish of criminal anthropology" according to Lombroso himself[77]—Thomson was arguing that crime was "bred in the bone" and noting "the ugliness and deformities of criminals, their under size and weight, and other evidences of degeneration." "On the border-land of Lunacy lie the criminal populations," he suggested.[78] The criminal class was marked by peculiar hereditary physical and mental characteristics that were allied to disorders of the mind. This class had a "*locale* and a community of their own" in the cities: "The greatest number are thieves, Ishmaelites, whose hand is against every civilized man. There is a thieves' quarter, a devil's den, for these city Arabs." Born into crime, "as well as reared, nurtured, and instructed in it," their criminal habits became "a new force, a second nature, superinduced upon their original moral depravity."[79] Invoking a medical metaphor, Thomson claimed that crime was "incurable . . . hereditary in the criminal class" and transmitted "like other hereditary maladies."[80] Thomson called for transportation to the colonies, the breaking up of criminal communities, and lengthy sentences for habitual criminals. He concluded on a melancholy note: "The criminal hereditary *caste* and character, if changeable, must be changed slowly, and how to do it must be to sociologists and philanthropists always a *questio vexata*, one of the most difficult state problems."[81]

The project of maintaining the boundary between savagery and civilization—a project shared by ethnographers and social analysts alike—remained a continuing source of unease for many Victorian commentators on crime.[82] Despite their belief in the malleability of character, even the phrenologists argued that a small number of "irredeemables" were beyond recovery due to deficient intellectual organs. It was certain, Henry Maudsley asserted, "that lunatics and criminals are as much manufactured articles as are steam engines and calico-printing machines, only the processes of the organic manufactory are so complex that we are not able to follow them."[83] Anyone who had studied the "step-children of nature"[84] recognized them to be "a distinct criminal class of beings, who herd together in our large cities in a thieves' quarter, giving themselves up to intemperance, rioting in debauchery, without regard to marriage ties or the bars of consanguinity, and propagating a criminal population of degenerate beings." It was "a matter of observation," Maudsley continued, "that this criminal class constitutes a degenerate or morbid variety of mankind, marked by peculiar low physical and mental characteristics. . . . An experienced detective officer or prison official could pick them out from any promiscuous assembly at church or market."[85] "There is a destiny made for a man by his ancestors," he wrote, "and no-one can elude, were he able to attempt it, the tyranny of his organization."[86]

Phrenology's radical faith in the malleability of human character helped to reorient criminal jurisprudence away from retribution and deterrence toward more systematic, proactive reformist principles. Its challenge to existing notions of criminal responsibility encouraged new ideas such as rehabilitation and reform, the proposal that the sentence must fit the criminal, and the concept of "criminal insanity." Phrenology also inspired prison administrators to conceive of penology as a science that might professionalize prison management.[87] By explaining criminality in terms of the mental faculties, phrenology promoted the idea that people varied in their propensity to offend. In endorsing the concept of crime-as-disease it profoundly influenced later approaches to criminality—such as those formulated by physicians and psychiatrists—that foregrounded degenerationist notions.[88] But above all, phrenology was important to the emergence of criminology, as it played a fundamental role in creating criminology's central figure of the born criminal type. By the 1870s, as empirical criminology was beginning to emerge, the idea of the born criminal had become an item of faith across a variety of disciplines. Cesare Lombroso—the man whom history records as the "father of criminology"—was far from unusual in insisting that the criminal was a special type of degenerate human being. Nevertheless, the energy that he devoted to studying and promoting the concept of "L'uomo delinquente" was remarkable.

One day, in 1911, in Mantellate jail in Italy, a twenty-eight-year-old inmate who was serving a two-year sentence for wife beating was summoned from his cell.[89] The peasant's obligation that day was to function as a case study for a class of police administrators. Professor Salvatore Ottolenghi examined the man, carefully measured his body parts, and pointed out certain features of his "antieurhythmic face." He also drew attention to the prisoner's receding forehead, his overly developed cheeks, and protruding bones. Ottolenghi further noted the inmate's scars, calluses, and lack of tattoos. He then questioned the prisoner about his childhood, family, health, moral habits, and criminal record. The professor concluded that the offender was "un tipo inferiore," a coarse criminal specimen. Here was a dangerous individual, the professor told his students, a man capable of committing violent crimes when under the influence of alcohol and when caught in a "morbid epileptic rage."

Ottolenghi was a dedicated follower of the man he referred to as "that titanic figure, Cesare Lombroso."[90] It had long been Lombroso's ambition to transform policing into "a scientific instrument . . . which employs photography, the telegraph, notices in newspapers, and above all knowledge of criminal man."[91] But it was Ottolenghi who did most to bring Lombroso's vision for "a scientific police that knows, with mathematical exactness, the physical characteristics of criminals" to fruition.[92] An intermediate-level bureaucrat, Ottolenghi transformed his mentor's unsystematic ideas into a new philosophy of policing and successfully introduced positivist criminology into government administration. The base for his operations was the School for Scientific Policing that he had established in Rome in 1902. He began publishing the *Bulletin of the School of Scientific Policing* in 1910. The school's extensive curriculum included the study of "Bertillionage" (criminal anthropometry), fingerprinting, photography, and criminal writings, as well as that of weapons, forged documents, and instruments for picking locks. Ottolenghi taught a course entitled "Applied Anthropology and Psychology" that instructed students how to recognize a criminal's "precise heredity, physical, ethnic, psychological, and pathological characteristics."[93] When the time eventually came, the ambitious Ottolenghi was anxious to seek Mussolini's support.

Lombroso's *L'uomo delinquente* of 1876 assimilated a number of threads of European thinking about crime into the single figure of the atavistic "born or instinctual criminal." Its author would later claim that this simple concept had come to him in "a flash of inspiration" while examining the unusual skull of Giuseppe Villella, a thief and arsonist. After having studied 383 skulls, Lombroso concluded that the criminal was characterized by an enlarged middle occipital fossa and vermis. He elaborated on these findings with an indefatigable series of anatomical, physiological, psychological, and moral tests, buttressing his argument with analogical correlations from nature and ethnological and linguistic studies. Criminal man was "an atavistic being who reproduces in his person the ferocious instincts of primitive humanity and the inferior animals."[94] Lombroso considered this unfortunate species of humanity a throwback to an earlier phase of evolution. Such criminals bore extensive signs of their degeneration on their bodies—apelike stigmata—that the expert eye could detect.

Lombroso had many intellectual debts. They included Gall's phrenology and Broca's craniology, the positivist philosophy of Comte, Haeckel's notion of recapitulation, and Spencer's psychology. As a medical student at the University of Pavia in the late 1850s, he had been impressed by the teratology and comparative anatomy of Bartolomeo Panizza. Morel's degeneration thesis and Marzolo's comparative linguistics were also important influences. Later on he drew on Darwin's ideas. Like the discipline he contributed so much to, Lombroso was an indefatigable assimilator. Although the idea of the born criminal was fundamental to his philosophy, the concept did not remain static across the various editions of *L'uomo delinquente*. The book went through five editions between 1876 (when it was two hundred and fifty pages long) and 1897 (when it consisted of three volumes of two thousand pages in all, together with an *Atlas* of illustrations). French and German editions appeared in 1887, and a short summary of Lombrosoian doctrine, *Criminal Man*, was translated into English in 1911.

Criminals were "constituted for evil" Lombroso wrote; they "do not resemble us, but instead ferocious beasts."[95] In the second edition of his book, he introduced the "habitual criminal," the "insane criminal," and the "criminal by passion." The "insane criminal" came to include three more psychological types, less distinguished by physical stigmata: the alcoholic, the hysteric, and the "mattoide" or semi-insane. Coined by one of Lombroso's most loyal followers, Enrico Ferri in 1880, the "born criminal" featured in the third edition of *L'uomo delinquente*. Lombroso's increasingly elastic categorization scheme resulted in the labeling of many more deviants as criminal.[96] He reduced the space devoted to biologically perverse criminals from a half to a third and expanded his discussions of the sociological causes of crime. Space allotted to punishment theory similarly increased across the editions. It was also in the third edition that he abandoned atavism (the reversion to a more primitive stage of evolution) for degeneration (the passage of pathologies through different generations of the same family), suggesting a prenatal mechanism for the latter mediated by the effects of alcohol, venereal disease, or malnutrition. Degeneration allowed Lombroso to incorporate many more stigmata and to integrate mental traits into his system. Despite gradually reducing his reliance on anthropometry and craniometry, by the fourth edition he claimed to have studied 6,608 criminals and, by the final edition, having measured 689 skulls.[97]

In the final edition of *L'uomo delinquente*, Lombroso added the category of the "occasional criminal," but complained that "it does not offer a homogeneous type like the born criminal and the criminal by passion, but is constituted of many disparate groups." Lombroso's followers were not troubled by the increasing flexibility of their leader's central concept; indeed, many of them produced their own variant on the criminal type. Enrico Ferri, for example, a brilliant criminal lawyer and Lombroso's "most visible and indefatigable disciple,"[98] proposed a scale of dangerousness that included the occasional criminal, criminals by passion, insane criminals, and—the most treacherous of all—the born criminal. Raffaele Garofalo, whose major work was *Criminology* (1885), posited three physiognomic types: the murderer, the violent criminal, and the thief. By 1900, the category of criminal man had expanded to include criminaloids, habitual criminals, criminals by passion, occasional criminals, and criminal crowds. The insane criminal had come to incorporate imbeciles, idiots, epileptics, the morally insane, manic-depressives, alcoholics, and the demented.

As criminological discourse expanded, so did the list of physical stigmata taken to signify deviance. Not only did asymmetry of the face, eye defects, and excessive jaw size count toward the diagnosis, but defects of the thorax, an imbalance of the hemispheres of the brain, and even the presence of supernumerary nipples were also taken to be indicators of criminal man.[99] Even before the publication of *L'uomo deliquente*, Lombroso had claimed that "as a rule," thieves had "mobile hands," rapists had "brilliant eyes" and "delicate faces," and murderers had dark, abundant curly hair.[100] The lack of agreement as to which stigmata signified criminality attracted much criticism.[101]

Despite the elasticity of the notion of the born criminal, there was a widespread belief among the faithful that the category represented a stable and special kind of contemptible human being. Garofalo concluded that "all who deal with the physical study of the criminal are forced to the conclusion that he is a being apart."[102] Despite accepting that tradition, prejudice, inadequate role models, climate, and alcohol were all implicated, he maintained that "there is always present in the instincts of the true criminal, a specific element which is congenital or inherited, or else acquired in early infancy and becomes inseparable from his psychic organism. There is no such thing as the 'casual' offender."[103] Boasting that his physiognomic theory rarely let him down, on one occasion Garofalo claimed that he had erred in distinguishing murderers from fraudsters "not more than seven or eight times out of a hundred."[104]

Criminology—a term apparently first coined in 1883—thus emerged as the systematic study of the peculiar biological abnormalities of criminal man.[105] That criminology was coupled to the criminal as one mountaineer to another was a fact recognized by contemporary observers. In 1894 T. S. Clouston wondered what "anatomical, physiological, and psychological signs are there to distinguish this criminal and his cortex?" "If there are no such signs then there is no such branch of science as criminal anthropology," he concluded.[106] Clouston recognized that the boundary between the born criminal and the habitual criminal was far from stable. But even if there was no "absolutely marked criminal type that all will agree on," there could be no doubt "that criminals fall far below a high or ideal anatomical and physiological standard of brain, and body and mind."[107] In a recapitulation reminiscent of Henry Mayhew's extensive studies of the London poor, Clouston argued that scientific criminal anthropology "must deal with the idle, the vagrant, the pauper, the prostitute, the drunkard, the imbecile, the epileptic, and the insane, as well as the criminal." This complex array could be reduced to "two great sources of criminality." First, "the not fully evolved man who might do his work well enough in a primitive society, but who cannot accommodate himself to the conditions of a highly organised and largely artificial modern society." And second, "the non-developed man, whose development has been pathologically arrested towards the end of the period of adolescence, just before the inhibitory and moral faculties had attained normal strength, there being in him often a slight intellectual impairment also."[108]

During the early years of the nineteenth century, phrenologists, statisticians, and asylum doctors had not been interested in singling out the criminal for special attention. They were more concerned with understanding the physical determinants of human conduct in naturalistic terms.[109] Their projects did not attempt to create a distinctive criminological science, even though many of their ideas and techniques eventually became part of criminological knowledge and practice. Nevertheless, these sciences, together with a variety of administrative projects, contributed to a general process that gradually made possible the systematic study of what J. Bruce Thomson called "a depraved and criminal class hereditarily disposed to crime." By the century's end, crime was no longer thought to be a mysterious occurrence, explicable only in terms of morality's failures. It had come to be regarded as an ordinary if regrettable feature of society, a natural phenomena whose regularities rendered it amenable to empirical investigation.

The explosion of international interest in criminal man left a considerable quantity of ideological debris in its wake.[110] At the 1889 Second International Criminal Anthropology Congress in Paris, French criminologists attacked Lombrosoian doctrine for its determinism, proposing instead to account for criminality with the concept of the "social milieu."[111] Gabriel Tarde and Paul Topinard voiced their opposition, as did Alexandre Lacassagne, who proposed a sociological explanation of crime.[112] Other challenges to the "Italian School" focused on the reality of cranial anomalies, the statistical data, and the lack of criminality in women.[113] The adoption of the Italian approach, Lacassagne had argued at the First Congress, would mean that jurists and legislators could "do nothing but cross their arms, or construct prisons in which to gather these misshapen creatures."[114] As the debates descended into disorganization and ad hominem attacks, the Italians voted en masse not to attend the Brussels congress. Criminal anthropology's detractors either took a sociological approach—the socialist Turati asserting that "bourgeois Society is the biggest criminal"—or they downgraded the importance of biology by pointing to the existence of habitual or occasional criminals. Colajanni, the author of *Criminal Sociology,* made the compelling criticism that Lombroso and his followers had failed to find a single trait that was exclusive to delinquents. The Catholic Church, a firm opponent of criminal anthropology, resolutely defended the concept of free will, rehearsing the argument that crime was a function of immorality.[115] Charles Féré, the author of *Degeneration and Criminality* (1895), blamed the genesis of crime on morality and inferior physiology. Charles Goring, a student of Karl Pearson's biometric statistical school, similarly conflated the physical, the mental, and the moral in his description of the criminal.[116]

French anthropology—a powerful influence by the final quarter of the nineteenth century—was more open to ethnographic and cultural interpretations of human action. Police officer and creator of the "Bertiollage" system of anthropometrics, Alphonse Bertillon refused to accept the notion of born criminal, arguing that attempts at rehabilitation would be useless.[117] One of Lombroso's most influential opponents, the sociologist Gabriel Tarde, disputed the theory's atavistic underpinnings, offering instead a psychosociological explanation of crime.[118] Tarde's social imitation theory proposed that criminal customs and habits were transformed into personological traits over time.[119] Hostile to the Italians' perceived fatalism, the French thought that heredity and social milieu were bound up in constant reciprocal exchange.[120] The "social milieu is the broth of criminality," Lacassagne wrote, "the microbe is the criminal."[121] Medical doctors maintained their influence within the French legal system. Even though they sought legal recognition for the limited mental responsibilities of the criminally insane, they attempted to avoid alienating jurists with excessive claims for institutional reform as the Italians had done.[122] Because the French critique was consistent with the concerns of the legal community, a split between the new ideas and the classical doctrines was avoided.[123]

Lombrosoian criminal anthropology spread like a virus across Europe. It was well established in Spain by the start of the twentieth century. Félix de Aramburu, the vice-rector of the University of Oviedo, had given the first public exposition of Lombroso's theories in 1886. Two years later Alvarez

Taladriz and Rafael Salillas at the University of Alava founded the *Revista de Antropología Criminal y Ciencias Médico-Legales*, a monthly periodical in imitation of Lombroso's journal. As the country's foremost representative and promoter of criminal anthropology, Salillas came to be known as Spain's "little Lombroso."[124] The penologist Pedro Dorado Montero and the lawyer and self-taught criminologist Bernaldo de Quirós y Pérez became the leading Spanish writers on the subject. The latter's *New Theories of Criminality* of 1898 gave prominent place to the innovators of the Italian school. By the early years of the twentieth century, the Italian positivist approach to crime had significantly permeated Spanish universities, the police and prisons, the Institute of Social Reform, parliament, and the government. As in Italy, criminal anthropology's mixture of radical and conservative tenets appealed to both ends of the political spectrum.[125]

From the 1890s into the early years of the twentieth century, both Italy and Spain were experiencing attacks from anarchist terrorists. Whereas in Italy criminal anthropology was a sufficiently influential doctrine to be used to placate and contain terrorism, this wasn't the case in Spain. In his 1894 book on anarchists, Lombroso had attempted to demonstrate that assassins and bombers "were epileptic, insane, the victims of congenital disease of various sorts, degenerate, hysterical, and often suicidal."[126] He reframed their politically motivated acts as the "deeds of the mentally unbalanced, juvenile delinquents, and common criminals." Thanks to "a virtual alliance" between the government and the Italian positivist school, criminal anthropology was able to obviate further dissent by individualizing and medicalizing political agitation.[127] The result was a reconceptualization of the meaning of violence and a refusal to create political martyrs who might foment further protest. The Spanish government's reliance on the classical notion of the rational and responsible individual law breaker led to a disastrous handling of two anarchist incidents in the first decade of the twentieth century.[128] In 1906, after a needlessly politicized trial for attempted regicide, the anarchist educator Francisco Ferrer was inadvertently turned into a heroic martyr by the authorities. Three years later, he was made a scapegoat for being the ringleader of a riot in which one hundred people were killed by army reservists. Whereas criminal anthropology might have pathologized Ferrer and weakened his influence, instead he "became another Giordano Bruno or Galileo, an enlightened thinker who seemed to be the innocent victim of the reactionary policies of a Spanish government dominated by clerics—the sacrificial offering of a new Spanish Inquisition."[129]

The German medical community showed a considerable interest in criminological questions from the 1880s. German doctors had long enjoyed a standing contact with the criminal justice system as forensic psychiatrists in the courts and as prison physicians in the correctional system. As a result, the medical profession felt compelled to respond to Lombroso's biological theory of crime.[130] By the 1870s it was standard procedure to call a medical doctor to the court if a defendant's mental condition was in question. Ten years later it was known that there was a greater incidence of mental illness among prison inmates compared to the general population, although it was difficult to distinguish those who became ill after conviction from those who were so beforehand. Whereas pre-Lombrosian psychiatry was concerned with the offender as an exceptional phenomenon, the exploration of a general link between insanity and crime became the norm after Lombroso.

The German reception of Lombroso's ideas took hold in the mid-1890s with expositions by Robert Sommer, Abraham Baer, Paul Näcke, Julius Koch, Hans Kurella, and Eugen Bleuler. Sommer accepted the notion of "endogenous criminal constitutions" but doubted the concept of "a type in the anatomical sense." Although they also denied the notion of the born criminal, Baer and Näcke proposed that an unfavorable social environment would trigger criminality among a degenerate subpopulation. Koch drew attention to "moral debility," a key factor in "cases in which immanent pathological characteristics of the individual turn a person into a criminal."[131] From the mid-1890s to the outbreak of the First World War, the most influential defense of the concept of the born criminal was in Emil Kraepelin's *Psychiatrie*, a textbook that was in its eighth edition by 1915. Kraepelin had first introduced Lombroso's work to the medical profession in his favorable review of *L'uomo delinquente*. Although he accepted that different types of criminal had differing somatic constitutions, Kraepelin rejected Lombroso's atavism hypothesis and considered the concept of stigmata unnecessary. Degeneration theory became highly influential in German psychiatry, partly because it could be used to explain virtually any mental illness through its positing of generational decline, and also because it was sufficiently flexible to accommodate both hereditarian and environmental aetiologies as well as explaining minor or borderline psychiatric conditions.[132] Kraepelin, Bleuler, and Koch's approach stripped the notion of the born criminal of its anthropological characteristics and redefined the concept in purely psychiatric terms. Gustav Aschaffenburg and Paul Näcke took a more complex view of the interaction of heredity and environment, arguing that many criminals suffered from general mental abnormalities that made them more likely to succumb to a life of crime. This approach subsequently came to dominate German criminology.[133]

In 1898 the Austrian judge Hans Gross founded the *Archiv für Kriminal-Anthropologie und Kriminalistik*. As the title of the journal suggested, the intention was to combine the study of scientific crime detection and the handling of scientific evidence. Gross had studied physics, psychology, medicine, and general science, as well as microscopy and photography.[134] Every criminal case was a scientific problem to which the examining judge must apply the best scientific and technical aids available. Gross rejected Lombroso's theory of the born criminal, maintaining instead that criminals functioned according to normal psychological mechanisms, knowledge of which was a crucial part of the investigating officer's armory. A Criminalistic Institute was established at the University of Graz in 1912, and Gross' *Manual for the Examining Justice* went through seven editions before 1915.

In continental Europe during the last quarter of the nineteenth century, criminology assembled itself around the concept of the "born criminal." A "relic of a vanished race,"[135] the born criminal was an anomaly, a prehistoric savage "living amidst the very flower of European civilization."[136] The notion of the born criminal was a rich and heterogeneous tapestry of concepts that weaved empirical data with the wisdom of folklore and tied the utopian dream of a crime-free state to an imaginative use of scientific technique. The new science resulted from transformations in an array of enterprises ranging from statistics, prison reform, and psychiatry.[137] A variety of administrative elements also found their way into the discipline, including charitable and social work, the management of workhouses and slum housing, inquiries about the causes and extent of inebriety, and investigations into the employment and treatment of children. Although all ended up as ingredients in the modern criminological mixture, at the time "they were discrete forms of knowledge, undertaken for a variety of different purposes, and forming elements within a variety of different discourses," none of which corresponded exactly with the criminological project that eventually formed.[138] Only when a form of inquiry emerged that centered on the criminal could these various enterprises be drawn together under the umbrella of a specialist discipline.[139]

In the early nineteenth century the practical skills of magistrates and police detectives was of crucial importance in detecting the false appearances of disguised professional criminals, but toward the end of the century science came to organize knowledge around the criminal. While the earlier regime had been intimate and personal, and dominated by the immediacy of hands and eyes, the latter was abstract and theoretical, and governed by the distancing effects of physiological instruments and statistical tables. Although the miscreant was initially conceptualized as a fallen man, a victim of his own lusts and lack of personal discipline, he was, nevertheless, thought to be capable of moral action. This was not the case for criminal man, who was considered an irredeemably degenerate being, one incapable of functioning adequately in the modern world.

A discourse of otherness par excellence, the new science of criminology spoke of such types as prehistoric humans and contemporary primitives, promiscuous women and delinquent children, epileptics, and the morally insane. Despite their differences, all these species of human beings were believed to share a genealogy with that archetypal but abject figure—a "distinct category of social perception and analysis"[140]—the born criminal, *homo criminalis*. The notion of the category dominated all discussions of criminality until after the turn of the century. But the belief in criminal man prevented the emergence of other biological approaches to criminality. Although the technology that would eventually make up the lie detector became available to criminology in the 1880s, the instrument's invention would have to wait until criminology had abandoned its first organizing concept of criminal man.

CHAPTER 2

"A vast plain under a flaming sky"
The Emergence of Criminology

> At the sight of that skull, I seemed to see all of a sudden, lighted up as a vast plain under a flaming sky, the problem of the nature of the criminal—an atavistic being who reproduces in his person the ferocious instincts of primitive humanity and the inferior animals. —Cesare Lombroso (1906)

Between the publication of Cesare Lombroso's *Criminal Man* in 1876 and Charles Goring's *The English Convict* in 1913, criminology emerged as a viable empirical endeavor. The period was a time of political strife across Europe, conducive to birth of an enterprise devoted to explaining and controlling deviance and dissent.[1] Like earlier phrenological and degenerationist discourses, also associated with political messages, criminal anthropology was compatible with socialism, fascism, and liberalism, not to mention racism, sexism, and imperialism.[2] Criminal anthropology blossomed together with a European-wide effort by progressive jurists and penal authorities to revise and update criminal codes that had been held since the late eighteenth century.[3] Most of the existing statutes did not recognize limited degrees of mental responsibility, and few prison systems distinguished between habitual and insane criminals. Despite its formal egalitarianism, by the 1850s, criminal justice was perceived, in Britain at least, as being particularistic, discretionary, and embodied in personal relations.[4] Criticism of the old system as a "lottery of justice" went hand-in-hand with appeals to base a new system on scientific (and, therefore, supposedly impartial) knowledge of the criminal.

A new human science, it has been suggested, does not emerge as a consequence of the accumulation and refinement of data and theory—these are science's ambitions once established—but rather as a function of widely held social anxieties.[5] A fledgling discipline obtains support if it addresses a moral panic or solves a social problem. Breaking with the Enlightenment principle that everyone should be treated equally, criminology promised to identify the sources of social danger by scientifically differentiating between intrinsic and extrinsic criminality.[6] Criminology constructed *homo criminalis* as the problem, offering itself as the solution. Constructed "at the crossroads of moral philosophy and everyday social policy,"[7] criminology was an empirical-political hybrid.[8] As with the medical, phrenological, and statistical sciences from which it emerged, criminology's scientific object was considered preeminently governable. Previously thought to be chaotic and unruly, criminological subjects were to be transformed into "calculable, disciplinable objects."[9]

The focus here is on power. The political dimensions of criminological discourse led to the functioning of the complex enterprise of criminology with relative coherence and stability. The discipline's language, including its argumentative and rhetorical strategies, as well as an interpretation of the role that "Lombroso" played within the discourse, were fundamental. Criminology endowed its texts and the image of its founding father with tremendous charismatic authority. Like the flywheel in a complex machine, charismatic authority ensured the smooth running of the criminological apparatus; indeed, this field has consistently invested in the charisma of its pioneers. The lie detector could not have been created without it.

In 1859, on the outbreak of the Italian wars of unification, Cesare Lombroso volunteered as a doctor in the army. He was particularly struck by his fellow soldiers' tattoos, wondering if they could distinguish "the honest soldier from his vicious comrade."[10] Between 1863 and 1872, he was a director of insane asylums. By 1876, when he was appointed to the University of Turin, he had already published on pellagra, cretinism, genius and insanity, brain pathology, and criminality. Italy was at this time in the throes of a debate about the definition of the nation state. According to some estimates, less than one percent of Italians spoke the national language, and seventy-five percent of the population was illiterate.[11] Lombroso's social evolutionary model of deviance articulated widespread social anxieties, proffered a new language of social representation, and created "a blueprint for disciplining groups that resisted integration into the new national culture."[12] His theory promised to unify

Italy's disparate cultures, languages, customs, and economies by delineating "a biological hierarchy that guaranteed power and control to white European adult men."[13] "Only we White people have reached the most perfect symmetry of bodily form," he wrote in 1871, the year of unification: "Only we have created true nationalism."[14] If the criminal could be understood scientifically, Lombroso argued, then he and other threats to the social order could be excluded politically.[15]

"One would have to be blind ten times over," Lombroso lamented in 1894, not to realize that "we are the second most backward amongst the peoples of Europe with regard to morality, wealth, education, industrial activity, agriculture, [and] justice." He added scornfully, Italy held "the first place when it comes to uncultivated, malarial land, endemic illness . . . crime and the weight of taxes."[16] Although the contemporary moral panic over crime seems not to have been underpinned by any statistical evidence of rising crime rates, there was, nevertheless, a widespread popular belief that crime was on the increase. Above all, Lombroso's ambition was to assist his nation's march toward modernity by defending the state against dangerous individuals, both from within and beyond the state's borders. Despite extensive emigration, this was a period of demographic and economic pressures in Italy.[17] Although the notion of an atavistic throwback found an enthusiastic audience in this climate, Lombroso's characterization of the Italian peasantry as savages was less a discovery as "a virtual reflex of the governing castes of the North when they ventured into the rural hinterlands."[18]

While the Italian criminologists made a significant and lasting impact on police science, their ideas met with considerable antagonism in the legislature.[19] The Lombrosian solutions to criminality—transportation, imprisonment, or elimination—contrasted with the morality-based punishments of the classical period, when crime had "a fixed exchange rate in punishment."[20] The Italian legal community was hostile to criminal anthropology, challenging the importance it attributed to human agency and criminal responsibility. Juries were disinclined to absorb scientific knowledge, resisting calls to repudiate their reliance on folk wisdom.[21] Raffaele Garofalo complained that judges and legislators continued to abstract the crime from the criminal. To the great disappointment of the positivists, the new Zanardelli Criminal Code of 1889 eventually became but a mere reworking of classical legal theory that privileged the concept of criminal responsibility via the doctrine of free will. The Italians' failure to influence their own criminal code was a sobering lesson for agitators elsewhere in Europe.[22] Positivist theories had the most impact outside the criminal code, notably in such bureaucratic contexts as police science and the prison system.[23] Lombroso despaired over what he considered to be Italy's backwardness compared to other European nations, which were implementing the reforms he championed, such as parole, criminal insane asylums, and youth reformatories. But although he did not live to see them, his ideas eventually left their mark on the Italian criminal justice system.[24]

Organized criminology was slow to develop in England.[25] Crime was not regarded as a revolutionary threat throughout the Victorian period, but it was considered "a nuisance, a puzzle and a social blight."[26] As in Italy, there was perception of a "crime wave" by midcentury, occasioned by fears over industrialization, urbanization and population growth, women's freedoms, and the relaxation of "moral restraint."[27] The view that habitual crime was getting worse was widespread during the first half of the century, although there was a decline in expressed anxieties about crime during the latter half of the century, particularly after the 1880s.[28] The Habitual Criminals Bill, passed in 1869, was a departure from the classical tradition of penology.[29] The success of the Victorian war on crime elevated the image of the new professional crime fighters in both fact and fiction, "but at the same time weakened the criminal image and diminished its moral meaning."[30]

The crowd was an alluring target for criminological doctrine. Constituted as "the apotheosis and the crisis of a science which had been from the beginning a project of social differentiation,"[31] the crowd provides a striking illustration of how criminology combined empirical, political, and moral claims. The French sociologist Gabriel Tarde, for example, claimed that criminal offenses were the "cutaneous eruptions of the social body; at times indices of a serious illness, they reveal the introduction, through contact with neighbors, of foreign ideas and needs in partial contradiction of national ideas and needs."[32] Constructed as a diseased, primitive, or childlike entity, the crowd was an emblem of the problems that beset mass society. Crowds were considered feminine, vulgar, and unruly until governed by a member of the rational male elite.[33] Crowds had to be controlled whether their origins were biological or social.

The modern world was a hazardous place. According to Edinburgh psychiatrist Thomas Clouston, an important cause of criminality—not to mention insanity—was the failure of some adolescents to adapt to the rapidly changing conditions of modern life, due to a failure to develop the necessary psychological capacities for self-restraint. "In the course of the development of the brain I think it is a certain fact that the later years of adolescence are those in which the great inhibitory, moral, and social faculties that fit men and womenkind to live in a well ordered, modern, civil society, attain such perfection as they are capable of in most men and women."[34] Because mental inhibition was "the colonel-in-chief of the brain hierarchy," a breakdown in its development would result in criminality.[35] In bridging the social and the biological, inhibition was a key conduit within discourses of control where brain sciences, medical practice, social policy, and civil and criminal law interacted with questions of personal conduct and morality. An individual's "power to exercise self-control seemed to make possible a legally ordered society."[36] Governing the criminal had become a psychological matter.

The criminal was to submit to punitive measures of discipline, incarceration, and segregation, because he or she was considered a dangerous and degenerate subclass. But he was also the focal point in discussing projects of rehabilitation and reform. In the early nineteenth century criminals were thought to benefit from confronting stark symbols of their moral failings, while toward the end they became magnets for new welfare agendas. Less in need of deterrence and discipline, criminals were believed to require direct therapeutic intervention.[37] The advance of naturalism in the human sciences encouraged interventionist approaches. Henry Mayhew expressed this paradox in 1862 when discussing what he called "that portion of our society not yet conformed to civilized habits." Not only were habits, reason, and the will fundamental aspects of human nature for Mayhew, but they were also ideal candidates for modification. He pointed out that "the greater number of criminals are found between the ages of 15 and 25; that is to say, at that time of life when the will is newly developed, and has not yet come to be guided and controlled by the dictates of reason." The period when human beings began to assert themselves was "the most trying time for every form of government—whether it be parental, political, or social; and those indomitable natures who cannot or will not brook ruling, then become heedless of all authority, and respect no law but their own."[38]

After what has been called an "efflorescence of scientific reinterpretation of criminality" from the later 1860s to the early 1870s, the advance of criminological naturalism stalled in England, just as Italian positivist criminology was in the ascendant.[39] When it did emerge, British criminology was, to a considerable extent, connected to both penal reform and medico-legal circles, unlike Lombroso's anthropology, which struggled to influence those domains. The reorganization of prisons after nationalization in 1877 left few opportunities for prison doctors to pursue investigations not connected to their statutory duties.[40] The practical situation did not encourage theorizing around *homo criminalis*, despite phrenology's influence and Francis Galton's attempt to capture the essence of criminal physiognomies with composite photography. The British were modest in their claims and "respectful of the requirements of institutional regimes and legal principles."[41] British psychiatry did not isolate discrete human "types" and classify them by means of racial and constitutional differences. A therapeutically oriented practice, it was based on classifying mental disorders that demarcated the condition separately from the sufferer. Although criminals could exhibit a variety of conditions including insanity, moral insanity, degeneracy, and feeblemindedness, in Britain they were not conceived of as a distinct psychological type.[42] As *The Belfast News-Letter* put it in 1898, "The gallery of English criminals proves what many know already—that in England there is no criminal type. Lombroso's gloomy work, that would ascribe all crime to hereditary taint, can scarcely be read patiently by those who know English criminals."[43]

Two models of the criminal emerged in England toward the end of the nineteenth century, neither particularly Lombrosoian. First, the traditional working-class criminal was considered to be in retreat, thanks to a perception of increased police presence on the streets and a conception of the criminal as not so much degenerate as enervated of vital energy. A second image of the criminal as a skilled professional was posited "chiefly by the spokesmen, real and fictional, of the new professionals charged with carrying on the war on crime."[44] After the 1880s, the English press regularly reported on Lombroso's ideas.[45] But there was never enthusiasm to organize criminology around the notion of the born criminal, and many of the main actors later modified their earlier claims concerning the criminal "species." Despite the enormous concern with degeneration and eugenics in England from the 1880s to the early 1900s, British criminology was linked only precariously to the eugenics movement.[46] Such notions were tied to deep-rooted fears about the city and its outcast but fecund populations.[47]

The most enthusiastic follower of Lombroso in England was Havelock Ellis, whose widely read book *The Criminal* (1890) summarized Lombrosoian doctrine. "In Great Britain alone during the last fifteen years," Ellis complained, "there is no scientific work in criminal anthropology to be recorded."[48] He did his best to promote the science—reviewing some nineteen works written in four different languages for the *Journal of Mental Science* in 1891 alone. But by then the British medico-legal profession was becoming somewhat dismissive

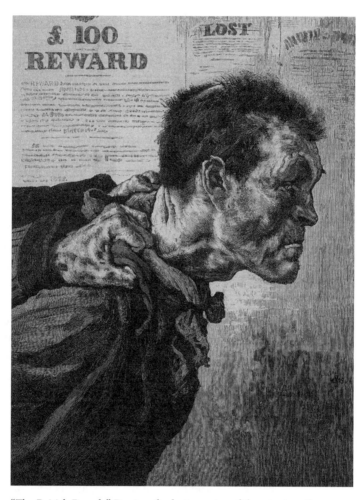

"The British Rough." During the last quarter of the nineteenth century criminology assembled itself around the concept of "the born criminal." *The Graphic* (London), Saturday, June 26, 1875.

of Lombroso's ideas.[49] In his presidential address to the 1894 British Medical Congress, for example, Dr. Long Fox briefly noted "The Great Subject of Criminology." The president joked, "that the human race is composed of two divisions of men—the criminals who have been found out and those that have not yet been found out."[50] Havelock Ellis hoped that the Rev. William Morrison's book, *Crime and its Causes*, would "do much to revive an interest in the scientific study of criminality in England," even though its author had devoted as much space to the environmental causes of crime as he had to the "humbly developed mental organization" of criminals.[51] Criminal anthropology was but a "science in embryo stage," asserted the Departmental Committee on Prisons in 1895, and most prisoners were "ordinary men and women amenable, more or less to all those influences which affect persons outside."[52] Morrison, representing H. M. Prison Wandsworth, attended the 1896 Geneva Congress of Criminal Anthropology, as did Galton and Arthur Griffiths. Griffiths wrote a report for the Home Office on his experience. A prolific author of both factual and fictional crime accounts, it was he who wrote the 1911 *Encyclopedia Britannica* entry on criminology, which was critical of type theory.[53]

In 1901 Dr. G. B. Griffiths, Deputy Medical Officer at Parkhurst Prison, inaugurated an anthropometric study to compare criminals with noncriminals. Some three thousand prisoners were measured. Commissioned and written under the auspices of the Home Office, the research was taken over by Charles Goring and published as *The English Convict: A Statistical Study* (1913). The book would later be depicted as the beginning of scientific criminology in England and as the final refutation of Lombroso's ideas. In line with his biometric orientation, Goring argued that the criminal should be viewed not as a qualitatively different morbid human kind—"no evidence has emerged confirming the existence of a physical criminal type"[54]—but rather as a variant of normality, differentiated only by degree. He pointed out that so-called criminal "anomalies" were only "more or less extreme degrees of character which in some degree are present in all men." Yet there was, he wrote, "a physical, mental, and moral type of normal person who tends to be convicted of crime."[55] Although Goring rejected Lombroso's degenerationist thesis, he retained the belief that the criminal had peculiar physiological characteristics. In the words of the eugenicist Leonard Darwin, criminals were "not a class apart, but merely ordinary individuals with certain innate qualities exceptionally well marked."[56] If criminality was a variety of normality then criminological science should be based on the careful measurement and statistical analysis of large populations. Such procedures would reveal meaningful patterns in the mass of data that would otherwise remain invisible.[57] By the First World War the idea of a separate criminal class had been replaced by an acceptance of criminality as related to the normal workings of society. Although he had dismissed Lombroso's notion of the criminal type, Goring, nevertheless, accepted in principle the idea that criminals could be subjected to eugenic measures on the basis of their low intelligence and poor physique.

Criminal anthropology fared much better in the United States. From the mid-1870s to the 1920s, biological theories of the born criminal affected public policy, encapsulating hereditary explanations of social problems.[58] "One nation ... America," Lombroso's daughter later triumphantly recalled, "gave a warm and sympathetic reception to the ideas of the Modern School."[59] Criminal anthropology did not meet with universal acceptance, but perceptions of social deterioration—thought to have been brought about by mass immigration—and beliefs about the progress of biological science created a favorable climate for its reception.[60] Many commentators expressed reservations about Lombroso's theory, but its emphasis on degeneration resonated well with widely circulating ideas about eugenics.[61]

Popularizations and translations of Lombroso's writings and Moritz Benedikt's *Anatomical Studies upon Brains of Criminals* (1881) introduced the science to an American audience. By arranging translations of Lombroso, Garofalo, and Ferri—while neglecting much of the French environmentalist school—the American Institute of Criminal Law and Criminology was an important conduit. Perhaps the most influential writer in English was Havelock Ellis, whose book *The Criminal* was in its fourth edition by 1911.[62] One of the earliest reports on the Italian school of criminology for an American readership came from the psychologist Joseph Jastrow.[63] Writing in *Science* in 1886, Jastrow reported that, according to the Italian scientists, crime was "the expression of a dangerous trait of character"; criminality, "a morbid phenomenon ... a defect analogous to insanity or idiocy." Jastrow wanted "to call attention to the fact that a change in our view of crime and criminals seems about to take place."[64] He lamented that while the law regarded chemical knowledge "as final" it considered medical knowledge to be provisional. The acceptance of Lombroso's theory of the insane criminal, however, would change the judiciary's guiding principles, he predicted. "Our knowledge of these marked classes is becoming sufficiently accurate and scientific," he concluded, "to warrant a practical application of these views in the legal trials."

In April 1891, Ronald Fletcher delivered his retiring presidential address to the Anthropological Society of Washington.[65] Claiming that nothing had yet been published in the United States on criminal anthropology, Fletcher's aim was to give his audience an impartial account of the work of the "New School of Criminal Anthropology." Conceding that the school was divided whether there was a definite criminal type, "a variety of the human species who has degenerated physically and morally," Fletcher quoted Lombroso's distinction between the murderer and the thief with approval. Fletcher's description of the murderer's "cold concentrated look" wouldn't have looked out of place in Bram Stoker's *Dracula,* which was still some six years away from publication: "Sometimes the eye appears injected with blood; the nose is often aquiline or hooked, always large; the ears are long; the jaws powerful; the cheek-bones widely separated; the hair is crisp and abundant; the canine teeth well developed, and the lips thin; often a nervous tic or contraction, upon one side of the face only, uncovers the canine teeth, producing the effect of a threatening look or a sardonic laugh." The thief, however, had "less cranial capacity than the Assassin," as well as "a remarkable mobility of countenance, the eye small and restless, the eye-brows thick and meeting, the nose flat, and the forehead always low and retreating."[66]

Fletcher was enthusiastic about the successful penal experiments at Elmira, the State Reformatory of New York.[67] Here, the inmates were taught trades and, proceeding on the assumption that "you cannot have a healthy mind without a healthy body," provided with a "good diet, athletic exercises, military training, an elaborate system of baths, massage, and other methods known as belonging to scientific gymnastics." Of 324 inmates paroled during the year, 148 went directly to employment at the trades they had learned in the reformatory. Despite having reformist ambitions for their charges, the prison physician, nevertheless, considered the typical inmate to be "undersized" and anatomically anomalous, "his weight being disproportioned to his height, with a tendency to flat-footedness. He is course [sic] in fiber and heavy in his movements, lacking anatomical symmetry and beauty. The head is markedly asymmetrical, with the facial lines coarse and hard, characteristic of a degenerative physiognomy."[68]

In 1892, D. G. Brinton explained to readers of *Science* that criminal anthropology was "one of the most actively cultivated and also one of the most immediately practical branches of anthropology." It consisted of three departments: observation, explanation, and application. While the first noted criminals' "anatomical and physiological peculiarities," the second attempted to explain such peculiarities by relying on "the laws of heredity, atavism, congenital tendencies, early impressions and pathological sequelae." Basing itself "on the inferences thus drawn," the third suggested "modifications in penal laws and the management of reform schools and houses of detention." So constituted, anthropologists considered the new science to be "the only method of procedure to deal intelligently with the great and growing problem of criminality."[69] In 1894 Brinton reported that the American legal profession was "almost unanimously" critical of the notion: "Take, they say, a dozen criminals as they come into the dock, wash and dress them as neatly, and they will certainly look as well as the dozen men in the jury box impanelled to pronounce upon their misdeeds."[70] It was surely the case that many men became criminals "through want, misery and destitution." But "as many more have not suffered in this manner; and a large class of crimes demand a well-regulated life for their commission." "Of course," Brinton concluded, "exception must be made in either case, of mental alienation, idiocy, insanity and the like." And therein lay the dilemma: despite sounding a note of skepticism about the "so-called 'criminal type,'" Brinton recognized that no one could deny the existence of psychopathology: criminality and insanity were intrinsically linked.[71]

This belief was widely held between 1890 and 1910—the years when criminal anthropology exerted its greatest influence. Arthur MacDonald's *Criminology* (1893) was the first work of criminal anthropology in the United States and Philip A. Parsons' *Responsibility for Crime* (1909) one of the last.[72] Phrenology had already encouraged Americans to link character and morality to biology, and Lamarckian and degenerationist notions had further correlated biology with social progress—and decline. This was a period in which both policing and penology were professionalizing and were receptive to new ideas that promised expert solutions to social problems. Criminal anthropology provided a useful set of explanatory tropes for a nation concerned with the illnesses, poverty, and biological dangers associated with a growing underclass. "The Degenerate Stock has three main branches, organically united," wrote social welfare worker Charles Henderson in 1893, "Dependents, Defectives and Delinquents. They are one blood."[73] In his *Diseases of Society*, Frank Lydston suggested that rapists should be castrated and all habitual criminals sterilized: "The confirmed criminal . . . is simply excrementitious matter that should not only be eliminated, but placed beyond the possibility of its contaminating the social body."[74] W. Duncan McKim's solution for a "tremendous reduction in the amount of crime" was even more extreme: "the *very* weak and the *very* vicious" should be afforded a *"gentle, painless death"* by gassing with carbonic acid.[75] In effect, the people "most concerned with crime control were receptive to the idea of the criminal as a biologically distinct and inferior being."[76] As had been the case in Italy, in the United States the impact of criminal man was to refigure the social problem of crime as biological.

Criminal anthropology was in a constant state of flux. It was not an ordered and precise system in which the criminal body was fully decipherable, but an unruly and often obscure discourse. Incapable of delineating criminal man's precise qualities, the science instead provided a flexible set of interpretative notions that were widely applicable. Lombroso's ideas appealed to the police and also to administrators of asylums, hospitals, and orphanages.[77] The fecundity of categories such as "moral insanity," "atavistic stigmata," and "inhibition" was a function of their plasticity and inherent ambiguity.[78] Above all it was the network of associations between physical, psychological, moral, and political domains that established scientific criminology. *Homo criminalis* might have been elusive but he was very promiscuous.

Criminal man was the offspring of a variety of heterogeneous discourses. Criminological texts were consequently a captivating assemblage of words, statistics, and images. Lombroso's description of the brigand Villella's skull for example—"the totem, the fetish of criminal anthropology"—might well have been a post hoc rationalization of events, but its legacy was a dramatic origin myth:

> At the sight of that skull, I seemed to see all of a sudden, lighted up as a vast plain under a flaming sky, the problem of the nature of the criminal—an atavistic being who reproduces in his person the ferocious instincts of primitive humanity and the inferior animals. Thus were explained anatomically the enormous jaws, high cheek bones, prominent superciliary arches, solitary lines in the palms, extreme size of the orbits, handle-shaped ears found in criminals, savages and apes, insensibility to pain, extremely acute sight, tattooing, excessive idleness, love of orgies, and the irresponsible craving of evil for its own sake, the desire not only to extinguish life in the victim, but to mutilate the corpse, tear its flesh and drink its blood.[79]

Considering the long history of the many concepts that went into making criminal man, it is telling that Lombroso claims to have discovered the essence of the criminal in an Archimedean moment of sudden insight. Although this passage contains some unusually arresting imagery, its poetic language was far from unique within criminal anthropology, a discourse that was often composed for a mass-market audience. Lombroso retained a strong interest in language throughout his career, a concern his biographer attributes to the influence of the linguist Paolo Marzolo.[80]

Aiming to communicate "the positivist gospel to the masses, criminal anthropologists were indefatigable in giving public lectures and writing articles for popular audiences."[81] Such an ambition required a sensitive ear for rhetoric and a keen awareness of the power of metaphor. In his 1893 book *Prisoners and Paupers,* the American criminal anthropologist Henry Boies, for example, wrote that criminals were "the imperfect, knotty, knurly, worm-eaten, half-rotten fruit of the human race."[82] Lombroso's own writings were bursting with quotations from literature, history, and folklore, all burdened with carrying the same evidentiary weight as personal observations, statistics, and experimental data. Aiming to be simultaneously popular and scientific, Lombroso was perpetually drawn to sensational examples that supported his argument.[83] Anecdotes were presented in captivating, lugubrious prose: "It is almost superfluous to record once again the instance of the aboriginal Australian, who, in reply to an inquiry as to the absence of old women in his country, said, 'We eat them all!' and on being remonstrated with for such treatment of his wives, answered, 'For one whom we lose, a thousand remain.'"[84] Criminal anthropology thrived on the sober reporting of the scandalous. Anecdotes that made the same point could be stacked one after the other in an apparent parody of inductivist science. Like so-called "savage races," criminals were idle. As the New Caledonians were stereotyped, they would rather die than work. North American Indians were thought to enjoy savage games: so did criminals. In the same way that South American Indians were thought to be incapable of blushing, so criminals were considered shameless. Criminals were inveterate thieves, a flaw they were thought to share with British New Guinea natives.[85] Criminal anthropology was obsessed with "otherness": the child, the woman, the so-called primitive, the mad. It was an otherness that had to be relentlessly reiterated.

In *The Female Offender,* amid the exhaustive presentation of prosaic statistics, Lombroso and Ferrero occasionally devoted space to particularly noteworthy case studies. Three photographs of "The Skull of Charlotte Corday," assassin of Jean-Paul Marat, were accompanied by a detailed description of her "extraordinary number of anomalies." Having discussed wrinkles—zygomatic, goniomental, and labial—in their chapter on the "Anthropometry of Female Criminals," the authors recalled "the proverbial wrinkles of witches, and the instance of the vile old woman, the so-called *Vecchia dell' Aceto* of Palermo, who poisoned so many persons simply for love of lucre." The authors revealed that although their evidence concerning the wrinkles had come from a photograph of a statue, it nevertheless possessed good evidentiary value of the criminal's "virile angularities." The bust, "so deeply wrinkled, with its satanic leer, suffices of itself to prove that the woman in question was born to do evil, and that, if one occasion to commit it had failed, she would have found others."[86]

In their chapter, "Vitality and Other Characteristics of Female Criminals,"

Lombroso and Ferrero presented a list of anecdotes featuring historical and classical figures in support of their claim that "if statistics are silent . . . history and tradition are there to show that the women who most frequently survive accidents and incidental and professional maladies are not the women of purest life."[87] The chapter "The Born Criminal" consists of little more than explanations of general principles interspersed with terse descriptions of notorious female villains, whose appalling crimes are described with staccato prose: "Tiburzio, after having killed a companion who was pregnant, bit her ferociously"; "Ta-ki used to order pregnant women to be torn limb from limb"; "Hoegli beat her daughter, and plunged her head into water to suffocate her cries"; "Pitcherel poisoned her neighbour out of revenge for his having refused to consent to his son's marriage"; "Jegado constantly poisoned people without any object"; "Sophie Gautier killed, by slow torture, seven children who had been given into her care"; "P . . . preferred to wound [her ex-lovers] by throwing into their eyes a powder made of fine glass which she had crushed with her teeth."[88] The wicked deeds of more than sixty named women are thus described, although some are used in evidence on three or four occasions at different points in the chapter. The narrative darts along, rarely dwelling long on any one particular theme. Single sentences are packed with disparate but colorful images: "It is a familiar remark in farmhouses that the most active and the readiest servant-girls are the least honest; while as for prostitutes, their agility is proved by the numbers among them who are dancers and tight-rope performers; and there is no cocotte who does not fence."[89]

Narrative was one of criminal anthropology's most important popularization devices.[90] Lombroso's 1902 study of the capture of the "celebrated brigand" Guiseppe Musolino for *Nuovo Antologia* magazine, for example, appealed both to conservatives and liberals due to its rich if logically inconsistent framework that attributed criminality to both innate physical factors and environmental forces. Guglielmo Ferrero deployed rhetorical stylistics to great effect in his account of the murderess Ernesta Bordoni for his *World of Crime* (1893), reporting the lurid facts of the case before proffering a scientific diagnosis. Giovanni Falco's account of one of Ottolenghi's case studies for the *Bulletin of the School of Scientific Policing* faithfully recorded the multiple voices of the participants, a tactic that encouraged the audience to make multiple interpretations.[91]

Narrative richness was part of the explanation why the criminal anthropologists became celebrated by Italian society.[92] Criminological texts were composed of anecdotes and also of stories culled from literature, philosophy, and linguistics, as well as facts from the animal and vegetable kingdoms. Lombroso was said to have observed that a vegetarian diet was less conducive to criminality than one based on meat.[93] An account of the effects of heat on plants could explain why crimes increased during the summer months. Never one to let a social trend pass without comment, he offered his thoughts on the role of the bicycle in crime and an "epidemic of kissing in America."[94] Ferrero argued "the females of the ants, bees, and spiders are particularly cruel because they are particularly intelligent." The "woman of to-day," he maintained, was "less criminal because less intelligent than the man."[95] According to Albrecht, there were numerous cases of slaughter and torture in the animal world, occasioned by the needs of defense (the bee), avarice (the ant), or sheer "love of fighting" (the cricket). Albrecht described how Lombroso found "bands of evil-doers among animals, fraud and thieving, and even criminal physiognomies..... the gray, bloodshot eye of the tiger and hyena, the hooked beak of the birds of prey, their large eye sockets and sexual perversity."[96] Havelock Ellis alerted his readers to research on antisocial conduct among rooks.[97] Criminal anthropology scrutinized the rodent-like cheek pouches of some criminals, the angle of the jaw of the lemur, and the supernumerary teeth of snakes. Reptiles, oxen, birds of prey, domestic fowl, and chimpanzees were also subjected to the physiognomic gaze. All this produced what has been appropriately called "a certain dizzying incoherence."[98]

In his *L'uomo di genio*, Lombroso derived an "index of genius" by analogy. Using a classification scheme based on geographical configurations such as mountains, hills, plains, and the nature of the soil, the criminologist deduced a correlation between genius and republicanism.[99] Analogical reasoning allowed criminology to make connections across time between social situations, political organizations, and disciplinary bodies. "Under certain unfavourable conditions," Lombroso-Ferrero wrote, such as cold and poor soil, "the common oak will develop characteristics of the oak of the Quaternary period. The dog left to run wild in the forest will in a few generations revert to the type of his original wolf-like progenitor." In humans, hunger, syphilis and trauma, and the abuse of drugs and alcohol, as well as "morbid conditions inherited from insane, criminal or diseased progenitors" could easily bring about "a return to the characteristics peculiar to primitive savages."[100]

Whereas criminal texts dating from the early nineteenth century are dominated primarily by words—particularly those uttered by criminals themselves—those published toward the end of the century commonly featured statistical tables, anthropometric measures, photographs of body parts, and illustrations of tattoos.[101] Such devices performed important rhetorical functions. The illustrations in Henry Boies' *Prisoners and Paupers* (1893), for example, consisted of photographs of immigrants at Ellis Island ("Typical Russian Jews," "A Group of Italians"), deformed "incorrigibles" at Elmira Reformatory, Roman statues, and a painting of a statesman. Criminology has been appropriately described as "an intertextual bricolage," a jumble of disparate elements drawn from various disciplines all devoted to persuading the reader that criminality was a part of nature.[102] Scientific criminology was, thus, elaborated from the familiar, its truth value "a coefficient of its relevance to its audience's needs and expectations."[103] Different types of evidence appealed to different audiences. The visual and verbal languages of criminal anthropology rendered the criminal body into an easily cognizable entity.[104]

This is not to say that criminal anthropology's rhetorical modalities were universally applauded. "One of the greatest defects of Lombrosoian presentation of criminal anthropological data," Harvard anthropologist and eugenicist Earnest A. Hooton later complained, "is the sensational anecdotal method which is used to clinch arguments."[105] Gustave Tarde had been even more critical: "What he calls experimentation—the accumulation of the mass of undigested remarks (absolutely sincere but uncritical) which he has since heaped together—has merely served to confirm him in his prepossession."[106] Yet it was precisely this jumbled anecdotal-analogical method that had helped to advance criminal anthropology from the beginning. Lombroso's texts are characterized by homologies between argumentative style, tabulated statistical data, and visual images that allow the reasoning to move effortlessly between hearsay, anecdote, and story. The statistics, such as they are, are usually nothing more complex than simple percentages, and their tabulated organization evidences no systematic rationale. The photographs of amassed criminal heads look—to modern eyes at least—like pages taken from a nineteenth-century *cartes de visite* album. All these disparate sources are piled up, one after the other, relentlessly aiming to persuade with an energetic display of sheer force.

Criminal anthropology conveyed information "efficiently, powerfully, and pleasurably."[107] With its habitual use of photographs of criminals and their skulls, illustrations of tattoos, and so on, criminal anthropology had enormous "visual clout." Both graphic and narrative forms of persuasion gave it enormous popular appeal compared to dry academic texts in other disciplines.[108] The science weaved together images of class, race, and gender with commonplace understandings of deviance to create an enterprise that had incredible epistemological power. At the heart of the project was the numinous figure of the born criminal: "lurid, horrifying, titillating, forbidden."[109] Such an exoticism was designed to evoke astonishment but not sympathy. Criminology had to appeal to a wide audience for financial and moral support. Yet it had to justify its status as a new discipline by constructing a unique and specialized scientific discourse. Unlike physics and economics, which had achieved the right to speak authoritatively about esoteric matters, criminology was obliged to amalgamate the traditional with the scientific.[110] Lombroso mobilized proverbs and folklore in support of his ideas, but he did not want to be perceived as having merely appropriated popular knowledge. This dilemma triggered new anxieties; if criminology was little more than a patchwork quilt of different ideas and practices appropriated from common sense, how could it claim to have special authority? And how could this "tension of expertise" be resolved?

The dilemma of expertise was resolved in part by the ways that "titanic figure," "the father of modern criminology," was depicted.[111] As early as 1869, a French traveler visiting Milan had described the occasion when Lombroso had discussed with him "certain anatomical indications by which criminals may be identified." The meeting left the traveler with the lasting impression that Lombroso was "a sort of monomaniac."[112] According to one scholar, "probably no name has been eulogised or attacked so much as that of Cesare Lombroso."[113] He was "a scientific Columbus who opened up a new field for exploration, and his insight into human nature was compared to that of Shakespeare and Dostoevsky."[114] Adalbert Albrecht conceded that because the science was "so intimately connected with the name of Lombroso ... no one can dispute his right to be considered the godfather of criminal anthropology." He was "one of the great men of the nineteenth century whose names were familiar to everyone, who were read by many but studied by comparatively few."[115] Max Nordau dedicated his *Degeneration* (1893) to his "dear and honoured Master." Yet as late as 1937, two criminologists felt compelled to point out that there was "no actual evidence in the voluminous criminological literature of the nineteenth century, before or after the time of Lombroso, which justifies the extravagant eulogies that are made of him."[116]

"The master" nevertheless attracted acolytes across Europe and the United States, many of whom took up his ideas with something approaching religious zeal.[117] In England, Havelock Ellis vigorously promoted his work. Dorado Montero translated Lombroso's writings into Spanish, taught the doctrines to his students—and was rewarded for his efforts with a lawsuit.[118] Spain's "little Lombroso," Rafael Salillas, became the country's foremost representative and promoter of criminal anthropology. In Germany, Hans Kurella promulgated the Italian criminologist's ideas in dozens of publications of his own and through translations of several of his works.[119] "His thoughts revolutionized our opinions," wrote Jules Dallemagne of Lombroso, and "provoked a salutary feeling everywhere, and a happy emulation in research of all kinds": "For 20 years, his thoughts fed discussions; the Italian master was the order of the day in all debates; his thoughts appeared as events. There was an extraordinary animation everywhere."[120]

That same force of character left a strong impression on Arthur Griffiths, who wrote a report on the fourth Congress of Criminal Anthropology in Geneva for the British government's Home Office: "Perhaps the most remarkable, and not the least interesting feature of the Congress was the tenacity with which Dr Lombroso held to his views. It is impossible to be brought into personal relations with this distinguished savant without being impressed by his sincerity, and the depth of his convictions. . . . Once in the course of the Congress, when very hardly pressed by certain hostile remarks, he cried, 'What do I care whether others are with me or against me? I believe in the type. It is my type; I discovered it; I believe in it and I always shall.'"[121] Lombroso was "a famous agitator," according to Gustave Tarde. He might have considered Lombroso to be a malign influence on criminology by the late 1890s, but he mobilized an appropriate metaphor in his grudging recognition of the Italian's powers: "We must acknowledge him to be the tongs to the fire by which we are consumed, but of the fuel which he added to it nothing remains except a handful of ashes."[122]

The unquenchable Lombroso wrote over thirty books and published one thousand articles, becoming one of the most prominent intellectuals in late nineteenth-century Italy.[123] He devoted enormous efforts to propagating his theory, disseminating his ideas in this continuous stream of publications.[124] Like the discipline he did so much to create, Lombroso had one foot in empirical science and the other in the emerging medium of mass culture. This was vital, because in order to win support, criminological theory had to leave the confines of the purely academic domain.[125] In this respect Lombroso was supremely competent. His sensational claims—criminals were "veritable savages in the midst of the brilliant European civilization"—appealed to a mass readership. His eclecticism and constant incorporation of new medical theories into the ever-expanding *L'uomo delinquente* assured him continued influence.[126] He was an expert assimilator of anecdotes and stories, empirical facts, and historical myths. In his 1895 *Contemporary Review* article, "Atavism and Evolution," for example, Lombroso touched on cannibalism, socialism, monarchy, ancient inventions, genius, brain organization, magic, medicine, and telepathy before concluding that "the same curvature which exists in the line of progress is also found in the course of retrogression."[127]

Lombroso clearly possessed a surfeit of what Max Weber called "charismatic authority," a character style unique to spiritual leaders, warrior heroes, and archetypal figures like the "sorcerer, the rainmaker, the medicine man."[128] At times of relative social and political calm, Weber suggested, bureaucrats and patriarchs usually administered the normal requirements of everyday life. Such periods favored leaders whose expertise was based on rational rules, and governance of the everyday was in the interests of social harmony. At times of social strife, however, periods of "psychic, physical, economic, ethical, religious [and] political distress," leaders were required that had a different array of "gifts of the body and spirit." Charismatic authority came to the fore during such periods of crisis.

Dismissive of organized economic structures and permanent arrangements, charismatic authority spurns salaries, promotion, and other established forms of career advancement. In its initial stages, it dismisses bureaucratic organization. The charismatic leader inspires his followers, demands their obedience, and insists on loyalty to an abstract ideal through the sheer force of his personality. Usually male—though not necessarily so—it is "the *duty* of those to whom he addresses his mission to recognize him as their charismatically qualified leader."[129] Those who triumph in trials of strength, perform incredible miracles, or undertake heroic journeys are rewarded with loyal followers. Lombroso was such a leader. He was perceived as a legendary hero, a visionary capable of seeing what no one else had seen: the diabolical glory of the born criminal. He was regarded as one of those rare individuals endowed with "supernatural, superhuman, or at least specifically exceptional powers or qualities."[130] According to Weber, the "god-like strength of the hero" makes "a sovereign break with all traditional or rational norms." Resolutely radical in his disavowal of preexisting laws and traditions, Lombroso possessed a single-minded rebellious streak. Paradoxically, his authority rested on an appeal to science, a form of instrumentalist organization par excellence.

Modernity, according to Weber, is characterized by the disenchantment with personal authority and its replacement by bureaucracy. But charismatic authority can nevertheless thrive in scientific domains, especially in situations replete with endemic uncertainty. Charismatic forms of authority are not premodern artifacts; they are central to the forces that shape late modernity.[131] Criminal anthropology was an ambiguous and conflicted scientific enterprise, one threatening to break apart under the weight of its contradictions. Lombroso's personal magnetism sustained the enterprise and defended it against hostile attacks. As an "inveterate dabbler," he could mediate between orthodox medicine and fringe practices, between science and superstition, and between the clinic and the theater.[132] He embodied criminal anthropology's contradictory tensions, but his status guaranteed the stability of the enterprise. He inspired great loyalty: "It has become a dogma that the dogma of the born criminal has been thoroughly disproved," wrote Zurich professor of psychiatry Eugen Bleuler in 1896, "but this is absolutely not the case. Not a single valid argument has been advanced against Lombroso's conception." It was an incontestable fact that criminals possessed specific "characterological attributes," according to Bleuler, such as "moral defects, a lack of inhibition [and] excessive drives."[133] Tarde's suggestion that the Lombrosoian Enrico Ferri was also able to reconcile opposites and balance opposing tensions could easily have been directed at the Italian master himself: "He possesses in the highest degree the faculty—natural to every born orator—of assembling and comprehending, in his ample formulas, the most contradictory assertions, without the slightest suspicion that there is any antagonism between them."[134]

The closer an empirical discipline gets to effective social organization, the more its claims become depicted as unified, essential, and unalterable.[135] Criminal anthropology, to a large extent, was held together by the force of Lombroso's personal magnetism. The science had to present itself as the compilation of undeniable truths in order to claim its right to define the problems associated with criminality and to recommend solutions.[136] Despite amassing a veritable encyclopedia of numbers, compelling images, and persuasive narratives, Lombroso never managed to define the criminal without ambiguity. Nor was this possible, even in theory, given that criminology was such an "intertextual bricolage," a mixture of discourses structurally incapable of producing unified knowledge. As late as 1897, Alfred Gautier, a Swiss professor, was still able to claim that criminology could be described "as a conglomerate, and not a science properly called."[137] "To what cause is this stationary agitation of criminal anthropology attributable?" Tarde asked. "I am bound to say, to the failure to get Lombrosianism out of the way."[138] Criminal anthropology was not moribund because "Lombrosianism is in its grave." "No!

The great hindrance to the progress of criminal anthropology is the obstinacy of its creator in adhering to the narrow conception, a hundred times demolished, which he formed in his youth, and to which he clings in spite of everything."[139] Tarde attributed the lack of progress in criminology to Lombroso himself. But what underpinned and stabilized criminal anthropology was the presence, practice, and persona of Cesare Lombroso. The charismatic figure, the image, the myth of Lombroso guaranteed the security of criminological discourse. A key principle of its sustainability and coherence, Lombroso's charismatic authority was one of the foundational elements of the spectacular science of criminology.

CHAPTER 3

"Supposing that Truth is a woman— what then?"

The Enigma of Female Criminality

> The significance of the factor of sexual overvaluation can be best studied in men, for their erotic life alone has become accessible to research. That of women—partly owing to the stunting effect of civilized conditions and partly owing to their conventional secretiveness and insincerity—is still veiled in an impenetrable obscurity. —Sigmund Freud (1905)

> Her great art is the lie. —Friedrich Nietzsche (1885)

One strand of the Western philosophical tradition has long considered woman to be an inherently secretive, deceptive, and duplicitous entity. At least since the Renaissance, the female body has been habitually associated with the mystery and generative power of nature. During the nineteenth century, women were increasingly defined by male scientists in terms of their corporeality: creatures enslaved by their passions.[1] Unlike the male mind, which was supposed to be unassailably free and governed by reason, the female mind was thought to be feeble, and easily overwhelmed and exhausted. The rational, public, trustworthy, and robust male mind was positioned at one pole on a continuum, the dangerous, private, deceptive, and delicate female body at the other. Empathy and insensitivity to pain were considered to be feminine qualities, to be generated and endured in private. Male emotions, however, should be governed by reason, particularly in public. Women should be confined to the domestic sphere where they could nurture children and tend to the emotional needs of their husbands. The home was thought to be where their dangerously unruly subjectivities could best be policed. It was believed that women should be governed, men should be governors.

As Tennyson put it in his poem, "The Princess" (1847): "Man with the head, and woman with the heart / Man to command, and woman to obey; / All else confusion."[2]

Central to the process and method of nineteenth-century science was an "eagerness to open up the woman and see deeply into the secrets of her body and of creation."[3] Late nineteenth-century criminology codified the female body as a cipher, a bearer of secrets about criminality. Women occupied a privileged position within criminological discourse.[4] It was within the female psyche that criminology sought out "illegible motivations" and "inscrutable sexualities."[5] The science came to regard the female body as a conundrum, nothing less than the numinous key to the puzzle of criminality. Like sexology and psychoanalysis, criminology was convinced that "the unknowable can be known, the enigmatic can be solved."[6]

Such ideas had a long history in the Western philosophical tradition. According to Aristotle, because a woman was a deficient male, her activities should be limited to the household. She should be denied a public voice or a civic role on the grounds that her nature required it. In the *Metaphysics,* Aristotle associated male with light, goodness, and the right; woman with darkness, evil, and the left. Plato's philosophy rested on a fundamental opposition between living a body-directed or a soul-directed life: the female body was considered to be the source of all the undesirable human traits.[7] The binary contrasts established by the ancient Greeks have had a pervasive influence on Western thought. According to Christian lore, God punished Eve's sins by imposing the curse of menstruation on her daughters. Menstrual blood was consequently considered shameful or contaminated, or to possess a "rank smell which any man could detect."[8] During the medieval period, nature was personified by two opposing female icons: an untamable temptress responsible for plagues and famines, and a serene, nurturing mother.[9] In the sixteenth century, according to the humoral theory of the temperaments, while men were considered hot and dry, women were thought to be cold and wet and therefore subject to "a changeable, deceptive, and tricky temperament."[10]

The ambiguous temptress / mother image was replaced during the scientific revolution of the seventeenth century by a system in which a dangerous and wild nature had to be tamed by mechanistic natural philosophy.[11] What was considered true of the natural realm was mirrored in the civic. In both France and England, for example, the legal subjection of wives to husbands was seen as guaranteeing the obedience of men and women to the slowly centralizing state.[12] English common law considered a married couple to be one person. A wife was deemed "covert," subsumed by her husband's legal personage. As a dependent and constrained person, it followed that a woman could not be trusted. Lying was considered vile, base, and mean because it arose from circumstances dictating the lives of ignoble people. As Montaigne put it, "children, common people, women, and the sick are most subject to being led by the ears."[13] It was the English gentleman's emancipated and unconstrained situation that was thought to guarantee the truth of his assertions.[14]

During in the eighteenth century—when "sex as we know it was invented" —the reproductive organs, for the first time, became the foundation of incommensurable differences between men and women.[15] Whereas women had been hitherto associated with the flesh, desire, and a willful sexuality, by the mid-nineteenth century ideologies of "maternal instincts" had made their appearance. This process was accelerated by the discovery, in 1843, of the involuntary periodicity of the reproductive cycle, which was interpreted as implying that female sexual pleasure was irrelevant to reproductive success. The Victorians separated the sexes into two radically different spheres that emphasized the dissimilarity of male and female bodies. The female body was associated with nature, passivity, emotionality, and irrationality; the male body with culture, activity, rationality, and reliability. Woman was thought to personify the object of knowledge, man the knowledge-making subject. "Woman is more closely related to nature than man," insisted Friedrich Nietzsche; "Culture is with her something external, a something which does not touch the kernel that is eternally faithful to Nature."[16] This historically situated dichotomy encouraged men to operate in the abstract and the public domain, while women were relegated to practical and domestic spheres. As G. W. F. Hegel put it in his *Philosophy of Right*, "the husband has his real essential life in the state, the sciences, and the like, in battle and in struggle with the outer world and with himself.... In the family the wife has her full substantive place, and in the feeling of family piety realizes her ethical disposition."[17]

Ascendant in the United States during the mid-nineteenth century, the "Cult of True Womanhood" prescribed "a female role bounded by kitchen and nursery, overlaid with piety and purity, and crowned with subservience."[18] Physicians advanced a view of women as frigid and advised them to embrace domesticity.[19] A social-political doctrine was, thus, systematically transposed into medical, scientific, and philosophical dogma. "To the man, the whole world was his world, his because he was male," lamented

Charlotte Perkins Gilman in 1911, "and the whole world of woman was the home because she was female."[20] Women were idealized as the guardians of domestic virtue. Body and mind were thought to be in harmony in women but conflicted in men. Men were obliged to be unemotional, their emotional and sexual impulses construed as being amenable to their will, indulged or repressed as necessary.[21] In private, however, men were permitted a broad range of emotional expression. Contrary to the stereotype of the emotionally limited Victorian male, the evidence suggests that middle- to upper-middle class masculine roles did not require men to be emotionally controlled and constricted at all times.[22] In the protected romantic sphere, it seems, men were permitted rich emotional lives.

"Woman seems to differ from man in mental disposition," wrote Darwin in his *Descent of Man* (1871), "chiefly in her greater tenderness and less selfishness." Bearing a heavy ideological load, Darwin's reduction of gender roles to biology was acute. Whereas a man engaged in competition with other men, displaying selfish ambition as a "natural and unfortunate birthright," a woman extended her "maternal instincts . . . towards her fellow-creatures." "It is generally admitted that with woman the powers of intuition, of rapid perception, and perhaps of imitation, are more strongly marked than in man," Darwin wrote, "but some, at least, of these faculties are characteristic of the lower races, and therefore of a past and lower state of civilization."[23] Most of the participants in these battles for social authority assumed and reinforced this binary model of difference articulated upon sex.[24] According to Auguste Comte, for example, women's status as the "emotional sex" consigned them to a "state of radical childhood."[25] For the neurologist George Beard, education could be hazardous to a woman's health because of the catalytic role it played in the aetiology of neurasthenia, the primary cause of which was "*modern civilization*, which is distinguished from the ancient by these five characteristics: steam-power, the periodical press, the telegraph, the sciences, and the mental activity of women."[26] In his *Principles of Sociology* (1876), Herbert Spencer argued that human advance depended on the expenditure of a fixed fund of "energy." The female body was considered to be at a disadvantage because reproduction diverted energy away from intellectual development.[27] In *The Evolution of Sex* (1889), Patrick Geddes and J. Arthur Thomson suggested that male cells were metabolically catabolic, aggressive, and active. Female cells, however, were anabolic, altruistic, and passive: "We have seen that a deep difference in constitution expresses itself in the distinction between male and female, whether these be physical or mental. . . . What was decided among the prehistoric Protozoa cannot be annulled by Act of Parliament."[28]

Evolutionary accounts of sexual differences were incorporated into highly prescriptive late Victorian gynecological and psychological theories.[29] The energeticist theory—taken up enthusiastically by Darwinian doctors such as Henry Maudsley and T. S. Clouston—was used to sanction opposition to any activity that threatened the cult of domesticity. The intense scrutiny devoted to the female body did not extend to the male body; men were defined by their rational deeds, not their sexual psychopathology.[30] To nineteenth-century medicine, those uniquely female aspects of a woman's physiology—menstruation, pregnancy, childbirth, lactation, and menopause—determined all of her other physical and social experiences. One physician in 1882 wrote that it was "as if the Almighty, in creating the female sex, had taken the uterus and built up a woman around it."[31] Many physicians accepted the thesis that to thwart the demands of the reproductive processes was to risk disease, insanity, and death.[32]

A corollary of the idea that women should be confined to the home was the claim that they were incapable of developing a sense of justice. Because they were thought not to be able to master the requisite capacities for participation in civic life, women were regarded as permanent sources of civic disorder.[33] For Rousseau (and later, Freud), whereas men could subdue and sublimate their sexual desires, women could not. Women had only the control of modesty, or as Rousseau put it, "a special sentiment of chasteness." They were creatures of passion, according to Rousseau, and must use "their natural skills of duplicity and dissemblance to maintain their modesty. In particular, they must always say 'no' even when they desire to say 'yes.' "[34] As Rousseau wrote in *Émile:* "Why do you consult their words when it is not their mouths that speak? . . . The lips always say 'No,' and rightly so; but the tone is not always the same, and that cannot lie. . . . Must her modesty condemn her to misery? Does she not require a means of indicating her inclinations without open expression?"[35]

Kant was scornful of the accomplishments of "learned ladies," asserting that "they might awaken a certain cold admiration by dint of rarity, but will at the same time weaken the charms which give them sway over the opposite sex."[36] Because women, according to Schopenhauer, lacked "reason and true morality," they represented "a kind of middle step between the child and the man, who is the true human being."[37] He pointed to their "instinctive treachery, and their irredeemable tendency to lying." Nature had apparently "armed woman with the power of deception for her protection."[38] Man, however, was endowed with a beard in order to help him disguise changes of expression when confronted by an adversary. Woman could dispense with this, because "for with her dissimulation and command of countenance are inborn."[39] Nietzsche also noted supposed female deceptiveness in his explanation of her disorderliness: "man wishes woman to be peaceable: but in fact woman is *essentially* unpeaceable, like the cat, however well she may have assumed the peaceable demeanour."[40] Nietzsche opened *Beyond Good and Evil* by asking: "Supposing that Truth is a woman: what then?" "But she does not *want* truth" he answered; "what is truth to a woman! From the very first nothing has been more alien, repugnant, more inimical to woman than truth—her great art is the lie, her supreme concern is appearance and beauty."[41]

They have "a mania for lying," Lombroso said of female hysterics. Hysteria shared many features with the discourse of female criminality.[42] The incidence of the most widely diagnosed female psychological disorder of the time peaked during the years 1870–1900, a period coterminous with criminal anthropology's own rise to prominence. Symptoms of "the quintessential neurosis, the primal pathology" varied widely but could include fainting, breathlessness, palpitations, sweating, dizziness, shaking, fits, paralysis, blindness, emotional outbursts, eating disorders, anesthesia, and convulsions.[43] One midcentury textbook listed "inflammation of the abdomen, loss of voice, hiccup, a peculiar cough, pain in a spot on the breast-bone opposed to one in the back, pressure on the two together being unbearable: pain in the left side, delirium of various kinds, and paralysis."[44] The women who were so afflicted were described by one physician as being possessive of "more than usual force and decision of character, of strong resolution, fearless to danger."[45] Intellectual and educated young women who sought advanced education or political representation were often diagnosed with the condition. Male doctors colluded with husbands and fathers in the diagnosis of wives and daughters, especially those who petitioned for divorce. Known as the "daughter's disease," hysteria has been interpreted as an embodied mode of protest for women deprived of social or intellectual opportunities.[46] Bourgeois Victorian women were caught in a double bind. They had to be feminine objects of courtship, delicate and innocent, and yet strong, pain-bearing, and self-sacrificing wives and mothers. The disorder's symptoms have, therefore, been viewed an expression of what was socially problematic for women at the time, a "psychosomatic protest" against suffocating social conditions. Mutism symbolized women's "lack of voice" in the wider society; anorexia echoed the "starvation" of women relegated to the domestic sphere.[47]

Both a product and an indictment of her culture, the hysteric was considered "duplicitous, self-dramatizing, and morally undisciplined."[48] Thought to be a common illness among prostitutes, masturbation was afforded a causal role in the disorder's aetiology. Only in the male could hysteria be considered a truly abnormal state, according to Thomas Laycock, who naturalized the hysteric as a "child-woman." It was but a short step from extreme nervous susceptibility to hysteria, and from there to overt insanity, claimed Dr. Isaac Ray in 1866. He proposed that pregnancy was a time when "strange thoughts, extraordinary feelings, unseasonable appetite, criminal impulses, may haunt a mind at other times innocent and pure." The bourgeois hysteric was considered idle, self-indulgent, and deceitful. "As a general rule," wrote the professor of clinical medicine at Marseille in 1883, "all women are hysterical and . . . every woman carries within her the seeds of hysteria. Hysteria, before being an illness, is a temperament, and what constitutes the temperament of a woman is rudimentary hysteria."[49] As one American churchman put it in the 1890s, "The excessive development of the emotional in her nervous system ingrafts on the female organisation a neurotic or hysterical condition which is the source of much of the female charm when it is kept within due restraint. In moments of excitement, it is liable to explode in violent paroxysms. Every woman carries this power of irregular, illogical and incongruous action and no-one can foretell when the explosion will come."[50]

Hysteria was considered nothing less than "a kind of pathological intensification of female nature itself."[51] An acute marker of the pathologizing discourse of the nineteenth-century female body, hysteria shared an explanatory framework with female criminality, whose features included biological determinism, sexuality, invisibility, emotionality, irrationality, pathology, and primitivism. Hysteria was considered such an intense exemplar of female nature that it was itself strongly suspected to be a form of malingering. According to Maudsley, hysterics were "perfect examples of the subtlest deceit, the most ingenious lying, the most diabolic cunning, in the service of vicious impulses." He denounced the "moral perversion" of young women who "believing or pretending that they cannot stand or walk, lie in bed . . . when all the while their only paralysis is paralysis of will."[52] "It is a most extraordinary fact, which it is of much importance to bear in mind," wrote the author of one midcentury medical textbook, "that there is scarcely a disease to which the human body is liable that is not simulated in hysteria, and it requires a most severe scrutiny to detect the fictitious and real malady.... Hysteric affection of the joints are very common, as also of the spine; and contractions of the limbs, obstinate vomiting, constipation, and even pregnancy, may be enumerated among the various simulated conditions."[53]

Writing in the *Lancet* in 1855, F. C. Skey argued that the doctor's first task was to distinguish the true illness from the false. "You would imagine this task an easy one, but it is not so. Diseases are feigned both willfully and unconsciously; the first are generally detectible by a discriminating judgment; the second are imitated by the hand of Nature herself, and are not so readily detected. This facetious condition of the body, that mocks the reality of truth,—that not only invades the localities, but imitates the symptoms of real diseases in all the diversity of its forms, deluding the judgment and discrimination of men of even considerable experience in their profession,— is known under the term *hysteria*."[54] "These patients are veritable actresses," complained Jules Falret, alienist at the Salpêtrière Hospital in 1890, "They do not know of a greater pleasure than to deceive.... In one word, the life of the hysteric is nothing but one perpetual falsehood; they affect the airs of piety and devotion and let themselves be taken for saints while at the same time secretly abandoning themselves to the most shameful actions."[55] The suspicion over the veracity of hysteria expressed the long-standing distrust of woman in general. In the same paradoxical way that the female hysteric was both suffering from a real illness and feigning an imaginary one, so the female criminal was saturated with an evident criminality yet exhibited no physical signs of that criminality whatsoever.

The notion that women were located in the middle of a hierarchy that placed "primitives," children, and the mad at the bottom with white European men at the top was widespread and long-lived. In 1875, W. L. Distant argued that the "form of the skull" and the weight of the brain demonstrated that "women hold an intermediate position between the child and the man." It was incontestable, he claimed, "that there are physiological conditions which must for ever tend against the possibility of women as a rule arriving at an equal, much less acquiring a superior, position to men in the mental struggle."[56] In 1899, Alfredo Niceforo pronounced that scientific tests had established that female traits were "impulsiveness, fickleness, puerile vanity, love for exterior appearance, and triviality, all the noted psychological attributes, in a word, that are common to women and savages."[57]

Having inaugurated the measurement and testing of women from the 1880s, criminology became increasingly obsessed with women, as evidenced by Lombroso and Ferrero's *The Female Offender* (1895), Salvatore Ottolenghi's *The Sensitivity of Women* (1896), and Scipio Sighele's *The Modern Eve* (1910). Ottolenghi's thesis was an elaboration of Lombroso and Ferrero's claim that "women . . . feel less as they think less."[58] Although the female offender was depicted as the mirror image of her male counterpart, she was nevertheless thought to possess some unique problematic qualities. Compared to a man, wrote Béla Földes, a woman is "more passive and less educated into controlling her will. She is often merely the intellectual factor, or the instigator, of crime."[59] A woman's culpability would be lessened if it could be shown that her biological state had been a causal factor in the execution of her offence. As Maudsley claimed, "cases have occurred in which women, under the influence of derangement of their special bodily functions, have been seized with an impulse, which they have or have not been able to resist, to kill or to set fire to property or to steal."[60] Whenever a woman has committed any offence, wrote Havelock Ellis, "it is essential that the relation of the act to her monthly cycle should be ascertained, as a matter of routine."[61]

Women might not be "given to intoxicants, nor to gambling, nor to roving," explained D. G. Brinton, but "they are more timid, more religious, more tender-hearted, and their sexuality is more passive." Brinton was reporting on Ferrero's recent "ungallant conclusion that the woman of to-day is less criminal because less intelligent than the man": "She is less disturbed by new ideas because she is slow to perceive them. When she is bad, however, she is 'very, very bad,' surpassing men in callous cruelty amid absence of pity or remorse."[62] Lombroso and Ferrero quoted Rykère's claim that feminine criminality was "more cynical, more depraved, and more terrible than the criminality of the male." They also approvingly quoted an Italian proverb that made the same point: "Rarely is a woman wicked, but when she is she surpasses the man."[63] They agreed with the prominent Italian folklorist Giuseppe Pitrè, who claimed that rarely "is the daughter of a bad woman honest, the son of a madman sane."[64] Another "terrible point of superiority in the female born criminal over the male," they concluded, "lies in the refined, diabolical cruelty with which she accomplishes her crime."[65]

Most of Lombroso and Fererro's *The Female Offender* was devoted to the discussion of pathology. A relentless encyclopedia of biological anomalies, the book contained chapters including "The Brains of Female Criminals," "Facial and Cephalic Anomalies of Female Criminals," and "The Criminal Type in Women and its Atavistic Origin." It is remarkable how much attention the authors devoted to physical measurements. In chapter 1 alone, for example, "The Skull of the Female Offender," the authors presented thirty-nine tables of various descriptive statistics, including measurements of "Cranial Capacity," "Facial Angle," "Bizygomatic Breadth," and "Weight of the Lower Jaw." In many cases the tabulations were ordered according to type of crime: homicide, infanticide, complicity in rape, prostitution, arson, and so on: "The frontal diameter is larger in prostitutes than criminals. Females guilty of rape and infanticide have the highest frontal index, and thieves and prostitutes the lowest. . . .The nasal index is inferior to the average of 48, especially among prostitutes, thieves, murderesses, and incendiaries."[66]

Industrialization, the changing birth rate, women's growing political emancipation, and other demographic and economic forces served to shift criminal anthropology's gaze toward women. The *Archives of Criminal Anthropology* regularly published articles on women in the 1890s, the decade that witnessed the emergence of first wave Italian feminism.[67] Ottolenghi acknowledged that the question whether "women feel more or less than men" was "intimately related to the difficult problem of the position of women in society."[68] Criminality, Lombroso and Ferrero had claimed: "increases among women with the march of civilization."[69] Because "housework and school work is presently improperly executed," Lombroso wrote in 1892, "our civilization has fallen in physical degeneration, pauperism and crime."[70] It was well known that in all countries there were fewer convictions for crimes committed by women compared to men. Contemporary European crime statistics apparently demonstrated that conviction rates varied "from the highest, 37 per cent, in Scotland, to the lowest, rather less than 6 per cent, in Italy."[71] Women made up ten to twenty percent of all offenders in Italy.[72] The upper figure is commensurate with the percentage of women convicted in England during the second half of the nineteenth century.[73] In England in 1870, women made up twenty-one percent of the average daily local prison population.[74] Yet in accordance with criminal anthropology's guiding degenerationist thesis, they were assumed to be more criminal—because less evolved—compared to men. This paradox was explained, first, by claiming that women exhibited less variation than men did, due to their reputed inherent conservatism. Their sex produced numerically fewer born offenders, it was conceded, but also fewer geniuses. Second, the statistics were assumed to underestimate true levels of female offending because "female crimes"—abortion, receiving stolen goods, and poisoning—were easily concealed.

Positivist criminology was convinced that female nature was innately duplicitous and secretive.[75] Women were considered to have a physiological incapacity for truth telling, and were thought to be habitual liars. This, it was claimed, was caused by the need to hide menstruation from men and sex from children.[76] It was due to "weakness, timidity, and shame of their sex," according to Béla Földes, that "women take greater pains than men to commit their crimes in such a way as to escape detection."[77] Földes echoed Quetelet's much earlier claim that women committed both fewer and different crimes compared to men because they were motivated more by the sentiments of shame and modesty.[78] "A further characteristic of female criminality is that, owing to the weakness and the sense of shame of those who perpetrate them, the crimes are less those of open violence than of deceit and secrecy."[79] In 1881, Vito Antonio Berardi claimed that the uterus holds "a tyrannical sway over the organism" so that it is called "the second brain of a woman."[80] For Guglielmo Ferrero, the female was characterized by deceit, dishonesty, a tendency for vendetta, and a passion for clothes. Ruled by an innate and eternal sexual force, even "normal" women were thought to be easily led into crime or prostitution.[81] Maternity, in contrast, was a "lofty affection . . . foreign to the degenerate nature of criminals," fostering a "moral antidote" to criminality.[82] The fixation on sexuality was the most significant and disturbing legacy of positivism to criminology.[83]

Fascinated by women in general, criminal anthropology's gaze intensified when it surveyed the prostitute. Lombroso's "elucidation of the connection between prostitution and all other kinds of crime will always remain a masterpiece," wrote one of his obituarists.[84] "The primitive woman was rarely a murderess," Lombroso and Ferrero claimed, "but she was always a prostitute, and such she remained until semi-civilised epochs."[85] By the time he was writing, prostitution had come to be regarded as the quintessential social evil, surpassing drunkenness, blasphemy, and adultery in the state's dossier of the undesirable.[86] In England, the prostitute was a prominent actor in the moral panic over the spread of venereal disease, which culminated in the three Contagious Diseases Acts of the 1860s. These legislated for the gynecological examination of any woman suspected by any policeman of prostitution.[87] "There is no woman, then, however virtuous," protested philanthropist and campaigner Josephine Butler, "to whom this law is not applicable, for there is no woman on whom the *suspicion of a policeman may not fall.*"[88] "Women criminals are almost always homely, if not repulsive," Mantegazza claimed, "many are masculine; have a large, ill-shaped mouth; small eyes; large, pointed nose, distant from the mouth; ears extended and irregularly planted."[89] According to Lombroso and Ferrero, not only was the prostitute's foot shorter and narrower than in normal women, but it was also shorter proportionately to the hand.[90] The "greater weight among prostitutes" was "confirmed by the notorious fact of the obesity of those who grow old in their vile trade, and who become positive monsters of adipose tissue."[91] Within criminological discourse, the prostitute did not represent womanhood's abject other but was rather an emblem of the inherent depravity of all women. The male was considered normal, the female abnormal, but the prostitute was corrupt.

Lombroso and Ferrero complained that the Italian authorities refused to let criminologists study actual convicts. But thanks to one Madame Tarnowsky, the authors had available to them a collection of photographs of Russian and French female criminals. They thought nothing of making an assessment merely on the basis of looking at them. "Louise C.," whose "habits were vagabond and unruly," "threw her dolls in the gutter, lifted up her skirts in the street." Although she was only nine years old, Lombroso and Ferrero concluded that "she offers the exact type of the born criminal." "Her physiognomy is Mongolian, her jaws and cheek-bones are immense; the frontal sinuses strong, the nose flat, with a prognathous under-jaw, asymmetry of face, and above all, precocity and virility of expression. She looks like a grown woman—nay, a man."[92] "Contrarily to criminals," Lombroso and Ferrero wrote, "these women are relatively, if not generally, beautiful." "Some of the photographs are quite pretty," they conceded: "This absence of ill-favourdness and want of typical criminal characteristics will militate with many against our contention that prostitutes are after all equivalents of criminals, and possess the same qualities in an exaggerated form. But in addition to the fact that true female criminals are much less ugly than their male companions, we have in prostitutes women of great youth, in whom the *beauté du diable*, with its freshness, plumpness, and absence of wrinkles, disguises and conceals the betraying anomalies."[93]

The lack of visible stigmata was, by no means, thought to be evidence of a lack of criminality. On the contrary, an absence of pathology was itself considered a cause for suspicion: "It is incontestable that female offenders seem almost normal when compared to the male criminal, with his wealth of anomalous features."[94] This puzzled Lombroso. It was an undoubted fact, he wrote, "that atavistically [the female] is nearer to her origin than the male, and ought consequently to abound more in anomalies." And yet "an extensive study of criminal women has shown us that all the degenerative signs . . . are lessened in them; they "seem to escape . . . from the atavistic laws of degeneration."[95] Lombroso and Ferrero suggested that prostitutes were obliged to hide their abnormalities with cosmetics and wigs. Furthermore, even if external anomalies were rare in prostitutes, internal ones, "such as overlapping teeth, a divided palate, &c.," were not.[96] Such disguises could not be maintained forever, however: "And when youth vanishes, the jaws, the cheek-bones, hidden by adipose tissue, emerge, salient angles stand out, and the face grows virile, uglier than a man's; wrinkles deepen into the likeness of scars, and the countenance, once attractive, exhibits the full degenerate type which early grace had concealed."[97] For positivist criminology, woman was a suspect category.

Lombroso and Ferrero proposed a number of theses in an attempt to explain "the rarity of the type" in women: sexual selection; the absence of stigmata "throughout the whole zoological scale"; women's "conservative tendency . . . in all questions of social order due to the immobility of the ovule compared to the zoosperm"; their lesser exposure to "the varying conditions of time and space in [the] environment" compared to men; and a less active cerebral cortex "particularly in the psychical centres."[98] The authors concluded that the crimes that women did specialize in, such as adultery, swindling, and prostitution, required an attractive appearance that prohibited "the development of repulsive facial characteristics."[99] Lombroso and Ferrero's preferred explanation for the lack of the criminal type in women, however, was atavism. Their account returned women to a period of human prehistory when there was supposedly less differentiation between the sexes. What was normal for men became criminal in women. Prostitutes were considered even less evolved than normal women, possessing characteristics such as small cranial capacities, narrow foreheads, left-handedness, and prehensile feet.[100] Although female born criminals were less common than the males, they were considered "more ferocious." Havelock Ellis preempted the objection that prostitution was more of a vice than a crime by asserting that from an "anthropological point of view" it was impossible to demarcate between the two forms of deviancy: "While criminal women correspond on the whole to the class of occasional criminals, in whom the brand of criminality is but faintly seen, prostitutes sometimes correspond more closely to the class of instinctive criminals. Thus their sensory obtuseness has been shown to be often extreme, and it is scarcely necessary to show that their psychical sensitiveness is equally obtuse."[101]

"Women's preference for strong scents," Albrecht asserted, "is to be explained only by the fact that they do not smell as keenly and therefore endure strong odors better."[102] Criminals were thought to have an inferior sensibility compared to law-abiding citizens. Lombroso invoked a hierarchy to conceptualize the problem, claiming that sensibility was highest in con artists but lowest in robbers. Criminals were more likely to be sensitive to the effects of metals and magnets and have acute eyesight. Garofalo pointed to the widespread practice of tattooing among criminals as evidence of their relative insensitivity to pain. Because "the normal woman is naturally less sensitive to pain," wrote Lombroso and Ferrero, women would possess less compassion— "the offspring of sensitiveness"—compared to men.[103] Women were also assumed to be less sensitive to pain compared to men because of the burdens of childbearing. Women's cruelty was a consequence not only of their weakness, according to Albrecht, but also of their higher pain thresholds. Other deleterious consequences followed from this: "Compared to that of men the morality of women is also inferior. They know only one honor, honor of sex. This inferior morality, too, comes from their lesser sensibility and intelligence, for also in the latter respect women are inferior. The highest plane of intelligence, genius, is completely lacking among women."[104] Francis Galton confirmed women's inferior sensory powers, claiming to have found that "as a rule that men have more delicate powers of discrimination than women": "The tuners of pianofortes are men, and so I understand are the tasters of tea and wine, the sorters of wool, and the like." He reasoned that if "the sensitivity of women were superior to that of men, the self-interest of merchants would lead to their being always employed; but as the reverse is the case, the opposite supposition is likely to be the true one." "Ladies rarely distinguish the merits of wine at the dinner-table," he concluded, and bizarrely, added that they were considered to be "far from successful makers of tea and coffee."[105]

Lombroso had conducted experiments on the "threshold of sensibility" to pain in 1867, using the electrodes of an induction coil. He and four male colleagues held the electrodes to various parts of their bodies, including their gums, nipples, tongues, lips, eyelids, soles of the feet, and the glans of the penis. Having previously used the device to deliver electrotherapy to his patients, Lombroso reported that the induction coil produced sensations such as a "series of hot pricks," "scalding pain," and pains like a "knife blade that passes through the joint."[106] The skin's thickness was proposed to mediate the experience of pain, which in turn was assumed to correlate with a person's intelligence. The mentally ill were alleged to be less sensitive to pain: "the demented, pellagroids, and apathetic melancholiacs presented diminished sensibility, erethismic melancholiacs presented increased sensibility."[107] The female body's alleged relative insensitivity to pain was in turn proposed as an explanation of why women had poor powers of sensory discrimination compared to men. Lombroso regarded most women as "frigid."[108] Salvatore Ottolenghi attributed women's insensitivity to pain to their increased vitality and longevity.

Physiological investigations—such as delivering electric shocks to the hand, tongue, nose, breasts, and female genitalia as in the so-called "Sensitivity Test"—was designed to empirically establish women's insensitivity to pain.[109] This research had some significant consequences. When their measurements were found to be greater than those taken from men, women were compared to children and were thus redefined as being "immature."[110] "What terrific criminals would children be if they had strong passions, muscular strength, and sufficient intelligence," asserted Lombroso and Ferrero, "and if, moreover, their evil tendencies were exasperated by a morbid psychical activity! And women are big children."[111] Women were thought to possess "many traits in common with children," including jealousy and a deficient moral sense. They were also considered "inclined to vengeances of a refined cruelty." Like children, female offenders were occasionally allowed to serve their sentences at home. The child was a "natural criminal," according to the recapitulationist logic. Women were also compared to sexually lascivious "savages." Women would "seek relief in evil deeds," the authors concluded, if their inherent "bad qualities" were intensified by "morbid activity of the psychical centres": "when piety and maternal sentiments are wanting, and in their place are strong passions and intensely erotic tendencies, much muscular strength and a superior intelligence for the conception and execution of evil, it is clear that the innocuous semi-criminal present in the normal woman must be transformed into a born criminal more terrible than any man."[112]

Having abandoned the search for physical stigmata of crime, criminology's investigation of female sensation led to a search for ordinarily invisible signs of crime within the body. From sensibility it was but a short step to sensitivity and from there to feelings and emotion. The study of blushing was a significant moment in the history of scientific study of emotion and a pivotal moment in criminology's turn to physiology. The history of shame in the nineteenth century begins with Thomas Henry Burgess's 1839 work, *The Physiology or Mechanism of Blushing*, which described blushing as a God-given "moral restraint," a mechanism to "control the individual from violating the laws of morality."[113] For Henry Mayhew, writing in 1862, while shame was "an educated sentiment . . . as *thoroughly* the result of training as is a sense of decency and even virtue . . . the main characteristic of civilized woman," it was "utterly unawakened in the ruder forms of female nature.... Many of the wretched girls seen in our jails have, we verily believe, never had the sentiment educated in them, living almost the same barbarous life as they would, had they been born in the interior of Africa."[114] Gender was inescapably entangled with race, class, and mental incapacity.

In her 1866 article, "Criminal Women," for the *Cornhill Magazine*, Mrs. M. E. Owen declared that a criminal man was "not so vile as a bad woman." Her anthropological metaphor was a bridge to Darwin's own reconceptualization of shame: "Women of this stamp are generally so bold and unblushing in crime, so indifferent to right and wrong, so lost to all sense of shame, so destitute of the instincts of womanhood, that they be more justly compared to wild beasts than to women.... Criminal women, as a class, are found to be more uncivilized than the savage, more degraded than the slave, less true to all natural and womanly instincts than the untutored squaw of a North American Indian tribe."[115] Although Darwin considered blushing to have been caused by shyness, shame, and modesty, his aim was to establish a continuity of emotion between man and animals. "Many a person has blushed intensely when accused of some crime," he wrote in *The Expression of the Emotions in Man and Animals*, "though completely innocent of it."[116] "A man reflecting on a crime committed in solitude, and stung by his conscience, does not blush; yet he will blush under the vivid recollection of a detected fault, or of one committed in the presence of others, the degree of blushing being closely related to the feeling of regard for those who have detected, witnessed, or suspected his fault."[117]

Although he conceptualized blushing as an involuntary biological mechanism, Darwin considered it to be a vestigial and useless artifact. For Lombroso, that criminals did not blush was an indication of their dangerousness; its absence signified "a dishonest and savage life."[118] Lacking normal affective capacities, their displays of inappropriate emotions such as pleasure at another person's suffering were further evidence of their degeneration. The inability to blush was also thought to be impaired among the insane. Havelock Ellis, Charles Féré, and G. E. Partridge regarded blushing as an "erethism" of sex and the origin of shame. "Inability to blush has always been considered the accompaniment of crime and shamelessness," wrote Ellis. "Blushing is also very rare among idiots and savages; the Spaniards used to say of the South American Indians: 'How can one trust men who do not know how to blush?'"[119]

According to Darwin, emotions had been produced by natural selection, and facial expressions of those emotions were remnants of once serviceable habits: "Every sudden emotion, including astonishment, quickens the action of the heart, and with it the respiration. Now we can breathe, as Gratiolet remarks and as appears to me to be the case, much more quietly through the open mouth than through the nostrils. Therefore, when we wish to listen intently to any sound, we either stop breathing, or breathe as quietly as possible, by opening our mouths, at the same time keeping our bodies motionless."[120] Darwin had been particularly impressed by the electrical experiments reported by Duchenne de Boulogne in his *The Mechanism of Human Facial Expression* (1862). Duchenne applied electrodes to different parts of the face in an attempt to simulate emotional expression.

What had made these investigations practicable was the prior historical construction of emotion as a biological phenomena.[121] The naturalization of the emotions made it possible to consider them as measurable entities, distinct from "the passions." By the mid-nineteenth century it was perfectly reasonable to think that emotions could be measured via the effect that they were reputed to have on the body. The origins of instrument-generated graphic representations of emotions can be traced to the mid-1860s, when the French physiologist Claude Bernard applied Étienne-Jules Marey's new cardiograph to record the heart's actions during emotional episodes.[122] Based on Ludwig's kymograph, Marey's sphygmograph was one of the first instruments to translate a subjective feeling into a graphical trace.[123] Bernard proposed that the slightest emotion produced a reflex impression in the heart that was "imperceptible to all, except for the physiologist" and his instrument.[124] Distinguishing between feigned and sincere emotions, Bernard suggested that only the latter would activate the involuntary physiological mechanisms necessary to produce a distinguishable and characteristic graphical recording.

A sustained program of research on the emotions began in the early 1880s when the Italian physiologist Angelo Mosso initiated a mechanistic, quantitative, and instrument-based approach to the study of emotions.[125] Mosso recorded the minute effects of induced emotions on laboratory animals and designed new instruments for determining the effects of emotion on the circulation. One of Mosso's experimental subjects was Michel Bertino, a thirty-seven-year-old man with a twenty mm gap in his skull.[126] Mosso's experimental technique involved rebuking Bertino while the registering apparatus inscribed blood volume changes in his brain. Mosso construed the resultant "cerebral autographs" as inscriptions produced by the brain: "See how the brain writes when it guides the pen itself."[127]

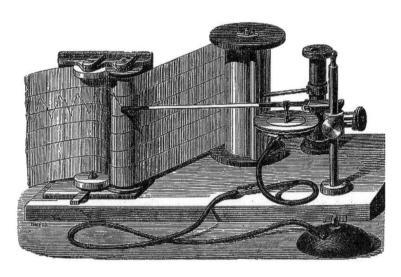

Étienne-Jules Marey's "polygraph taken out of its box and provided with the exploration device in a shell for the pulsation of the heart." From Jules É. Marey, *Mémoire sur la pulsation du coeur* (1875), p. 33, fig. 16.

Mosso's work appealed to Lombroso because it had the potential to render visible the criminal's dangerousness. Particularly important was the instrument's promise to record even "those emotions that are not depicted on the face."[128] Criminology enthusiastically embraced physiological instruments, because by illuminating the dark recesses of the body they promised to undertake the work of demarcating criminals from the insane.[129] Tamburini proposed that a "true psychometer" could enable comparisons between the mentally ill, and quantify the degree of alteration of the principal nervous centers. The sphygmic curves of the mad were believed to have a characteristic shape. Because the insane were thought to have less marked vascular reactions compared to intelligent subjects, instruments might aid in the detection of feigned mental illness. Lombroso thought that the instruments might be able to identify the particular physiological states that might enable an individual to commit a particular criminal act.

Criminology's authority came to depend on measurement devices. A well-appointed laboratory might list among its stock the following instruments: baristesiometer, campimeter, clinometer, craniometer, dynamometer, ergograph, esthesiometer, goniometer, Hipp's chronoscope, olfactometer, the *Schlitteninductorium,* spirometer, tachyanthropometer, thermesthesiome-

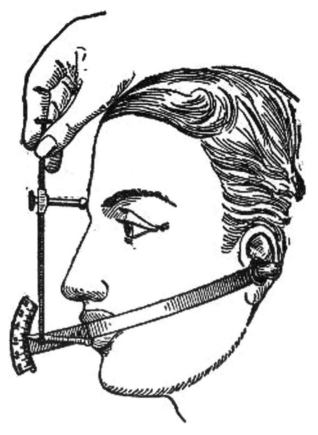

Arthur MacDonald's goniometer, used to measure the angle of the face. From Arthur MacDonald, "Psycho-physical and Anthropometrical Instruments of Precision in the Laboratory of the Bureau of Education" (1898), chap. 5, p. 1189, fig. 84. In: *The Experimental Study of Children*, 1141–1204 (United States of America, Bureau of Education, Report of the Commissioner of Education, 1897–98).

ter.[130] In spite of all this hardware, the turn to instrumentation did not produce the definitive empirical results the criminologists had hoped. But scientific instruments were not just expressions of the extension of the criminological gaze into hitherto unseen spaces; they were also tools for the fashioning of a scientific identity for criminal anthropology. These specialized techniques for reading the body furnished criminal anthropology with a "corporeal literacy that made possible an exegesis and a diagnosis."[131] An ability to manipulate scientific instruments and to gather data systematically was a crucial aspect of the construction and maintenance of scientific authority.[132] Like criminological texts, which presented table after table of data and heaped anecdote upon anecdote, criminology laboratories also amassed large quantities of fetishized scientific instruments. Lombroso's descriptions of esoteric scientific instruments were not restricted to his scientific texts; they also occupied a privileged place in his more explicitly popular texts. Readers were given instructions on how to make their own instruments and make investigations with them. The tachyanthropometer, for example, was designed, as Lombroso put it, to make "the practice of anthropometry very easy, even to people who are entire strangers to the science."[133]

Many of the experiments performed with these instruments had a theatrical dimension. A series of experiments in the early 1880s had multiple recidivists attached to a sphygmograph (on their left arms) and an induction coil (on their right) with a view to determining if electric shocks produced changes in blood pressure. Subjects were asked to listen to music or were shown images of nude women or money to see if such experiences produced physiological changes.[134] Criminals were asked to perform mathematical feats or were told bad news: "Are you aware that your mother is seriously ill?" On occasion the reactions of criminals were greater than those of normal subjects. Lombroso was fascinated by the "almost total lack of vaso-motor reactions" of some of his criminal subjects. One such, Ausano, was characterized as "prognathous, tattooed, receding forehead, with marked frontal sinuses, criminal uncle, drunkard father, neurotic mother; thief since childhood... never demonstrated any reaction, neither to music, a pistol shot, unpleasant news, nor calculations; only [an image of] wine produced a slight rise [in blood pressure] for 18 pulses."[135]

Another criminal, the recidivist thief Agagliate, demonstrated no response to mental arithmetic and showed only a slight reaction when shown a revolver. Thinking about the electrical shock generator merely "flattened the pulse, to render it almost barely detectable at the apices."[136] In 1885, American psychologist Joseph Jastrow reported in *Science* magazine that the study "of the bodily or anthropometric peculiarities of these defective classes is getting to be a very favorite line of work." "The sexual passion and the glass of wine modified the course of the pulse," wrote Jastrow, "the money did so still more, and vanity the most. In short, criminals are a very vain class of beings."[137] The aim of these interrogations—superficially similar to those that would be asked by lie detector operatives forty years later—was not to elicit informa-

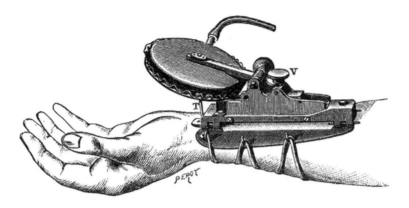

Transmission sphygmograph, 1882. All the lie detector's component parts had been invented prior to the beginning of the twentieth century. Charles Verdin, *Catalogue des instruments de précision servant en physiologie et en médecine construits par Charles Verdin* (Chateauroux, 1882).

tion but rather to see if the questions caused the subject "emotions of any kind, whether he has real affection for those beings to whom normal persons are attached, but towards whom born criminals and the insane in general do not manifest love."[138]

One effect of the use of instruments that would later feature in lie detector discourse was the way in which they de-emphasized the operator. The scientist "who was elsewhere so insistent on his special ability to read the body," was hidden from view.[139] This machine-operative tension was particularly marked for graphical instruments such as cardiographs, ergographs, pneumographs, and myographs. As Étienne-Jules Marey wrote in 1885, "Not only are these instruments sometimes destined to replace the observer, and in such circumstances to carry out their role with an incontestable superiority, but they also have their own domain when nothing can replace them. When the eye ceases to see, the ear to hear, touch to feel, or indeed when our senses give deceptive appearances, these instruments are like new senses of astonishing precision."[140] According to Claude Bernard, the cardiograph was "a veritable living machine." Its purpose was to yield a record unaffected by the prejudices of the observer, to bypass the ambiguities of language and speech altogether. As the body of the scientist receded, that of the suspect came to the fore, "imagined to tell its own truth."[141]

Enrico Ferri considered the sphygmomanometer to be useful "in the diagnosis of simulated disease" such as sham epilepsy. Ferri thought the instrument could also assist in the identification of criminals, like "tattooing, anthropometry, physiognomy, physical and mental conditions, records of sensibility, reflex activity, vaso-motor reactions, the range of sight, the data of criminal statistics."[142] Instead of alighting on a single reliable indicator of criminal pathology, criminological discourse once again searched for a definitive answer to the question: "What is a criminal?" The multiplication of instruments devoted to rendering the invisible visible mirrored the multiplication of physical stigmata and also vexed the criminologists. Junior researchers complained about the proliferation of instruments and the difficulties in getting them to work. In detailing criminal anthropology's encyclopedic intensity, Gabriel Tarde's eloquence betrayed a decidedly unpositivist poetic sensibility:

> Every instrument for measuring or which was known to contemporary medical science and to psychophysics, the sphygmograph, dynamometer, aestheseometer, etc., had already been employed by Lombroso for the purpose of characterizing in the language of figures or of graphic curves, singular arabesques, the manner in which thieves or assassins breathe, in which their blood flows, their heart beats, their senses operate, their muscles contract, and their feeling is given expression, and by this means to discover through all the corporal manifestations of their being, considered as so many living hieroglyphics to be translated, even through their handwriting and their signatures submitted to a graphical analysis, the secret of their being and of their life. In this manner he had discovered, especially by means of three sphygmographic tracings, that malefactors are very responsive to the sight of a gold coin or of a good glass of wine, and much less to the sight of a "donna nuda," in a photograph to be sure.[143]

The list was ostensibly impressive but its superfecundity evidenced an anxiety deep in the heart of criminology.

Lombroso used instruments like the plethysmograph and the "hydrosphygmograph" to confirm the absence of emotion in the habitual offender. When examining a criminal, Lombroso argued, it was essential to ascertain "whether he has any real affection for those beings to whom normal persons are attached, but towards whom born criminals and the insane in general do not manifest love."[144] The "Volumetric Glove," for example, tightly sealed, filled with air, and attached to Marey's tympanum, was used to ascertain if a person was a criminal on this basis: "If after talking to the patient on indifferent subjects, the examiner suddenly mentions persons, friends, or relatives, who interest him and cause him a certain amount of emotion, the curve registered on the revolving cylinder suddenly drops and rises rapidly, thus proving that he possesses natural affections. If, on the other hand, when alluding to relatives and their illnesses, or vice-versa, no corresponding movement is registered on the cylinder, it may be assumed that the patient does not possess much affection."[145] In 1893 Arthur MacDonald described how Mosso's plethysmograph could examine the "sensibility of criminals."[146] The instrument was used to investigate a criminal's "nervous and physical nature," his "mental depression," or the ability of children to perform mental calculations. It did not detect the lie. The instrument was assumed to be irrelevant for studying criminals' "power of deception," which was described under a separate heading in the article.[147] In 1899, Lombroso wrote that Mosso's plethysmograph was able "without affecting the health and without any pain, to penetrate into the most secret recesses of the mind of the criminal."[148] The target of this work was the personological criminal type, not the lie.[149]

"As a double exception," Lombroso and Ferrero wrote, "the criminal woman is consequently a monster."[150] The female offender was considered the exception that proved the rule precisely because she exhibited no degenerate stigmata. Whereas male criminals were believed to be pathological deviations from the normal, female criminals were considered invisible specters, "monstrous in their lack of deviation from the normal: they are a monstrosity in terms of criminal anthropology, the mythical creature at the heart of its labyrinth."[151] Criminology transformed the normal woman into an entity even more dangerous than the born criminal. The result was the barely legible *potential* dangerousness of the normal woman, the effect of which was to construct woman as "both normal in her pathology, and pathological in her normality."[152]

Criminology was one of many nineteenth-century discourses "determined to rend any and all veils of inscrutability, to bring women's secrets into the light of masculine inspection by whatever means necessary."[153] At the heart of the enterprise lay the mysterious enigma of woman, the carrier, embodiment, and essence of criminality.[154] By the end of the nineteenth century many criminologists believed that only by solving the enigma of woman could criminality be confronted. Criminology's elevation of women as the chief conundrum for their discipline was a stark example of the ability of the human sciences to assimilate long-held cultural beliefs and widely articulated social values into a systematic empirical framework. But the positioning of woman within late nineteenth-century criminological discourse was an important episode in the history of the lie detector for two reasons. First, the shift of attention toward the female body was accompanied by a move toward the investigation of the apparently inscrutable domain of the emotions, a domain that physiological instruments, it was hoped, would render visible. Second, as the central enigma of criminology, thought to possess an intrinsic but invisible guilt, the female body would provide the lie detector with an opportunity to become an indispensable new tool. One of the conditions for the lie detector's emergence was, therefore, the conceptualization of criminal woman as the quintessential problem of criminology. Another, however, would be the death of criminal man.

CHAPTER 4

"Fearful errors lurk in our nuptial couches"
The Critique of Criminal Anthropology

> It is on our domestic hearths that we are taught to look for the incredible. A mystery sleeps in our cradles; fearful errors lurk in our nuptial couches; fiends sit down with us at table; our innocent-looking garden-walks hold the secret of treacherous murders; and our servants take £20 a year from us for the sake of having us at their mercy. —Alfred Austin (1879)

The lie detector could not have been invented in the nineteenth century. The first obstacle was that the lie had yet to be constructed as an unambiguously shameful act as well as a psychophysical property of the normal human body—an entity whose expression should occasion guilt in the ordinary person. The Victorians were animated by the polarity of truth and deceit, but lying was emphatically not construed purely as an attribute of the untrustworthy and vulgar classes.[1] On the contrary, polite sectors of Victorian society considered the management of lying a valuable skill. Deception was a vehicle for socio-cultural transformation, while deceit generated certain forms of authority. A means of empowerment and for constructing selfhood, lying was culturally licensed, especially among the emerging class of bourgeois public intellectuals. The novelist Wilkie Collins, for example, attempted to incorporate dishonesty into the realm of the morally acceptable. Even toward the end of the nineteenth century the simple truth was only ambiguously attached to respectability. For criminologists, however, lying remained a cardinal trait of *homo criminalis*. Because the lie detector would come to rely on the lie as a feature of normality, the instrument's invention was not thinkable until criminology had abandoned the born criminal.

Toward the end of the nineteenth century anxiety was widespread that crime was not being removed but rather displaced, but not back to the slums of the criminal underclass. On the contrary, the naturalization of crime produced the alarming possibility that criminality might lie within the heart of the character of the ordinary law-abiding citizen. The public was concerned that certain crimes—particularly poisonings, embezzlement, and fraud—might not be coming to light, these crimes apparently committed by the respectable middle classes whose implementation was facilitated by the semblance of propriety. The perceived decline of Mayhew's riotous and menacing underworld made public life appear more secure. But uncomfortable questions were raised about the corruptions concealed within respectable society.[2] The perception that street-level crime was on the wane directed middle-class attention from the unruly masses toward the reputable citizen. Hitherto hidden forms of criminality became sources of anxiety.[3] The brutal murder of a small child at an English country house in 1860 horrified the nation, in part because it initially appeared that the killer must have been a member of the upstanding household. The search for the perpetrator of the "Road Hill House murder" inspired a national "detective-fever" and encouraged hundreds of people to write to the newspapers, the Home Secretary, and Scotland Yard with their proposed solutions.[4] "It is doubtful if public excitement ever rose to such a pitch in connection with a crime as that occasioned by the celebrated Road Murder," wrote the *Sporting Times* nearly thirty years after the event.[5]

The nineteenth century came to speak of crime and its punishments to a degree that was inversely proportional to their occurrences. Although gruesome punishment was gradually removed as a public spectacle, this led to an overproduction of images and narratives concerning crime and punishment.[6] As the nineteenth century progressed, the administration of the law was professionalized and individual testimony gave way to tangible evidence—embodied in the figures of the lawyer and the fictional detective.[7] Invisible threats appeared more frightening than perceptible, calculable ones. Novelists drew inspiration from the professional classes, the domestic family, and from their knowledge of morally ambiguous individuals. Charles Dickens and Wilkie Collins followed the Road Hill House case intensely. A safer society had been created at the cost of a need for greater vigilance. A growing preoccupation with blackmail highlighted new concerns about the concealment of secrets.[8]

The first fictional genre to exploit such anxieties was the "sensation novel" that began with Charles Dickens' *Bleak House* (1853) and *Great Expectations* (1860), and Wilkie Collins' *Woman in White* (1860) and *The Moonstone* (1868). By turning melodramatic conflict inward, the sensation novel outraged critics for depicting heroes as morally ambiguous figures. Paranoid and suspicious, and set within domestic realms, the sensation novel typically featured a reputable citizen—often a woman—gradually revealed to have criminal attributes. "In these days of fiction," wrote Alfred Austin in his 1870 essay, "The Sensation School," "a change has really come over the spirit of our dreams." "It is on our domestic hearths that we are taught to look for the incredible. A mystery sleeps in our cradles; fearful errors lurk in our nuptial couches; fiends sit down with us at table; our innocent-looking garden-walks hold the secret of treacherous murders; and our servants take £20 a year from us for the sake of having us at their mercy."[9] The effect of sensation was to make the familiar strange. By "domesticating the wild" the genre also "gothicized the normal."[10] The apparently normal fictional femme fatale, ostensibly a woman of good character, was an embodiment of criminality.[11]

Toward the end of the century, cheap production processes and rising literacy levels led to the explosive expansion of the market for novels and the popular periodical press. Novelists and journalists alike were seduced by criminology's rich imagery and explanatory principles. Detective fiction, after all, shares criminology's axiomatic premise: the identity of the criminal must become known, predicted, and recognized.[12] Founded in France in 1825, the *Gazette des Tribunaux* filed reports on sensational criminal cases and became a source of information for both scientists and novelists. Criminal anthropology's key tropes became widely known in realms beyond the scholarly journal and the academic congress. In its coverage of the infamous "Jack the Ripper" murders in London, for example, the press justified the promotion of order by mobilizing the figure of criminal man within the milieu of working-class Whitechapel. The demarcation between the normal and the pathological was further disturbed by the suggestion that "mad Jack" might not be a member of the underworld, but a respectable medical doctor, in view of his apparent knowledge of human anatomy.[13]

The press appealed to advertisers by reporting on outrageous stories that stimulated their readership; circulation increased with the promotion of populist causes such as nationalism, militarism, and imperialism. Spy and crime stories were also well received.[14] The press was motivated by an ideology of neutrality, impartiality, and exhaustiveness, and it aimed for full coverage of the "facts," in which both sides of the story were ostensibly reported. Focus was on individual criminal cases rather than on the complex social dimensions of crime.[15] The press reinforced social norms by privileging crimes of the lower orders, illegal escapades, stories of extraordinary misdeeds, and the daring exploits of master criminals. To protect the norms of the power interests that underwrote the press, criminal cases were discussed without referring to underlying socio-political conditions. It was imperative to titillate yet protect the reader with generic stories concerning cunning swindlers and their hapless victims. Thrilling but ultimately trivial crimes were reported in a terse, telegraphic fashion. Trials were narrated in the style of a stage play complete with star actors, captivating lawyers, and dramatic plots.[16]

The boundary between fact and fiction was porous. Crime fiction shared key images and tropes with journalism and criminology. Novelists contributed to criminology's project, promoting yet challenging the discipline's claims. In France, J. -H. Rosny aîné's *Dans les rues* depicted Fauburg street gangs as cavemen. Its Lamarckian thesis was that bad habits could turn a man into a criminal beast. Adrien Sixte, the central character of Paul Bourget's *Le Disciple* (1889), had been modeled on Hippolyte Taine, "the one the English gladly call the French Spencer." The story concerned the psychologist's attempts to discover criminal man. Emile Zola's *La bête humaine* (1889–90) had also been inspired by the Italian criminologist. (In a circular irony, Lombroso himself drew inspiration from the novel.) The story accepts the demarcation of criminal from civilized man as a social and biological fact, but order is achieved through an integration of criminal instincts rather than their negation, suggesting that everyone was potentially contaminated with criminality. Zola's work has been interpreted as dramatizing the contradictions and disintegration of the positivism that had initially inspired his project.[17]

Novelists such as Oscar Wilde, Robert Louis Stevenson, and particularly Bram Stoker all shared a fascination with the science of crime and the physiognomy of degeneration.[18] In Robert Louis Stevenson's *The Strange Case of Dr Jekyll and Mr Hyde* (1886), a Victorian gentleman-scientist regresses to an apelike primitive. In Rudyard Kipling's short story "The Mark of the Beast" (1890), an Englishman in India defiles a Hindu temple and regresses to a wolflike state. Although the story offered a rational, medical explanation for the degenerative transformation, it suggested that Western science was impotent to elucidate the mysteries of the East.[19] Decline had become a national concern following the formulation of the second law of thermodynamics in

1851. The discovery of the entropic cosmos occasioned reflections on the loss of human energies and the dissipation of vigor.[20] Richard Dugdale's 1877 bestselling account, *"The Jukes": A Study in Crime, Pauperism, Disease and Heredity* became a synonym for degeneracy on both sides of the Atlantic.[21] Scientific and literary texts alike expressed fears about dissipation, degeneration, and decline. The fin-de-siècle British Gothic novel was dedicated to exploring the boundary between the human and what has been called the "abhuman." A highly innovative genre that reemerges at periods of cultural stress, the Gothic violently enacts the reconstitution of the human body and mind. Scientists and novelists expressed nostalgia for the fully human subject, a project their work had undermined.[22] Themes of human devolution occurred repeatedly in the fin-de-siècle Gothic, showing the beast within or as a means of demonstrating how casually cruel a nature driven by random change could be.[23] The human body was conceptualized as utterly chaotic, unable to maintain its distinctions from an exuberant menagerie of spectacular possibilities: "slug-men, snake-women, ape-men, beast-people, octopus-seal-men, beetle-women, dog-men, fungus-people."[24]

An array of scientific discourses—including evolutionism, criminal anthropology, degeneration theory, sexology, mesmerism, and pre-Freudian psychology—all articulated new models of the abhuman body's ambiguities. With this crisis in the epistemology of human identity, science was infected with Gothic themes. Darwin had reconceptualized nature as a motiveless force, indifferent to suffering, that was, nevertheless, full of picturesque superabundance. Excessive and gratuitous, nature no longer evidenced divine benevolence but was a teeming chaos of meaninglessness and superfecundity. Natural selection incorporated the teratological and the fantastic, and gestured toward the grotesque.[25] Lamarckian evolution similarly provided fruitful explanatory tropes and expressed deep anxieties concerning the health of the nation.[26] The effects of alcohol, tobacco, opium, hashish, a poor diet, syphilis, and tuberculosis, as well as fears about the pathologies of the city were all seen as injurious to biological descent. Heredity was no longer considered the motivator of progress, but an invisible source of contamination and danger. In attempting to normalize the meanings of sexuality by delineating the many "perversions" the human body was capable of, sexology merely succeeded in multiplying sexual variety and enlarging the field of possibilities. It also undermined commonplace logics that posited corporeality as the exclusive property of femininity, a state that only masculinity could transcend. Gender was itself an unstable category that had to be perpetually fashioned anew as part of "the immense cultural labour" required to produce and sustain the liberal humanist subject.[27]

By critiquing science's classificatory schemes, the Gothic delighted in the creation of monsters. Literary narratives exposed the social, economic, and political conditions of knowledge production, questioned the ontological status of criminal man, and disputed the possibility of science ever purifying its categories. Profound questioning of truths and assumptions were found in the entertaining narratives of novels that revealed "the unsayable of social discourse."[28] By blurring boundaries, novelists were able to challenge the axioms of determinist science. The figure of the criminal "beast" became, on this reading, an intertextual species that slipped between science, literature, and the mass media. Bram Stoker's *Dracula* (1897), a masterpiece of ambiguity and anxiety, was heavily influenced by Lombroso: "The criminal always works at one crime [Van Helsing explained]—that is the true criminal who seems predestinate to crime, and who will of none other. This criminal has not full man-brain. . . . The Count is a criminal and of criminal type. Nordau and Lombroso would so classify him, and qua criminal he is of imperfectly formed mind."[29] The criminal tainting of aristocratic blood recapitulated the plotlines of several late Victorian novels.[30] Descended from "noble blood", the Count was, nevertheless, inherently corrupt. Animated by themes of hypnosis, hysteria, degeneration, and sexuality—a veritable Foucauldtian nightmare of biopower—the novel expressed the dangers of entering new places and experiencing new forms of consciousness. The poison of degeneration could be passed in blood from person to person, potentially infecting the entire population. Against the fantasies of Lombrosoians degeneration could not be contained. Deploying diaries, reports, and letters, the novel resists an easy synthesis. Yet it is concerned with resistance, frustration, and the failure of insight, "paralysed at a threshold of uncertainty, at the turning point between a psychiatric positivism (which the novel derided), and the glimpsed possibility of a new exploration of the unconscious."[31] Orthodox medicine is sleep walking, Stoker suggests, stumbling along in a half-light.

The modern figure of the detective emerged in the shadowy borderlands between vice and virtue, purity and corruption, and science and superstition. Arthur Conan Doyle's *A Study in Scarlet* (1887) has been credited with establishing many conventions of the detective novel such as scientific deduction as a means to knowledge. Holmes is nothing less than "the most perfect reasoning and observing machine that the world has seen."[32] As an admiring

Watson tells the great detective, "You have brought detection as near an exact science as it ever will be brought in this world."³³ Viewed simply as a detective story, *The Hound of the Baskervilles* appears to dispel magic and mystery and to render everything visible to the scientific gaze. Yet it is as much a fin-de-siècle Gothic tale as it is detective story. The novel was influenced by Lombroso, as Watson's representation of Selden demonstrates: "Over the rocks in the crevice in which the candle burnt, there was thrust out an evil yellow face, a terrible animal face, all seamed and scored with vile passions. Foul with mire, with a bristling beard, and hung with matted hair, it might well have belonged to one of those old savages who dwelt on the burrows in the hillsides. The light beneath him was reflected in his small, cunning eyes, which peered fiercely to right and left through the darkness, like a crafty and savage animal who has heard the steps of the hunters."³⁴ Relying on the conventions of contemporary Gothic narratives, it challenged the aspirations of science to subject crime and criminality to scientific analysis.³⁵ With its use of multiple narratives that implicitly critique science's claim to a single unified truth, the Gothic was a counterattack against the excessive faith in positivist science and a subversion of received wisdom about criminality.³⁶ Whereas Holmes— "part social physician, part magician"—can deploy his genius to solve crimes, Watson, a medical doctor, can solve nothing.³⁷ The novel is simultaneously a detective story promoting the exclusive use of reason and deduction and a warning about the failures of modernity and the futility of scientism.³⁸

Embodying the conflicting genres of the novel, Sherlock Holmes is associated with both light and darkness, with the urbanity and culture of modern London and with the elemental peat and granite of the primeval moor. "It is my belief, founded upon my experience," Holmes tells Watson in *The Copper Beeches* (1892), "that the lowest and vilest alleys in London do not present a more dreadful record of sin than does the smiling and beautiful countryside."³⁹ The paradox of Holmes is also the paradox of the detective story: the more it aims at a rational explanation of crime, the more its appropriation of the Gothic themes of criminal anthropology subverts that project.⁴⁰ The figure of the "criminal genius" highlighted the instability of criminal anthropology as it combined the normal and the pathological.⁴¹ "A criminal strain ran in his blood," Holmes says of Moriarty, a one-time professor of mathematics, "which, instead of being modified, was increased and rendered infinitely more dangerous by his extraordinary mental powers."⁴² Recall that for Lombroso, genius had a "degenerative character." For Lombroso, as for Galton, genius was a form of abnormality, a species of moral insanity. The hereditary line need not run straight; a "good birth" guaranteed nothing. In *The Man of Genius,* Lombroso warned, "even habitually sober parents, who at the moment of conception are in a temporary state of drunkenness, beget children who are epileptic or paralytic, idiotic or insane.... Thus a single embrace, given in a moment of drunkenness, may be fatal to an entire generation."[43]

Criminal anthropology therefore shored up the integrity of the liberal bourgeois subject just as it was undermining it.[44] The question of criminal genius was but one exemplar of this general social crisis of criminal nature and human agency. Articulating the oppositional, the unsaid, and the unsayable, literature challenged received scientific wisdoms. Robert Louis Stevenson's *The Strange Case of Dr Jekyll and Mr Hyde* (1886) disputed criminology's maxim that there was a qualitative difference between the normal and the criminal: "man is not truly one, but truly two." The novel suggested that within the healthy were hidden the seeds of the pathological: everyone was potentially suspect, a criminal in waiting. Inside everyman lay a dormant ailment, all too easily animated given the right circumstances. "I hazard the guess," Jekyll says, "that man will be ultimately known for a mere polity of multifarious, incongruous and independent denizens."[45] It has been suggested that Jekyll's moral and temperamental transformation was absent from the first draft of the story. Even in the final version, Jekyll had been living two lives for many years prior to Hyde's physical materialization. The scientist's experiment is less a fatal error than it is the polarization of the antagonistic tendencies of his nature.[46] In the end the novel, a thorough deconstruction of the Victorian cult of character, provides no solution to the dilemma as to how to expunge criminality from the self.

Physical decline and biological degeneration were an intoxicating mix, masterfully articulated by H. G. Wells in *The Time Machine* (1894). The future earth is imagined to be populated by two degenerate humanoid races, one lacking the power of reason, the other dissipated of energy. The Eloi were vapid and effete, their mental powers atrophied; the Morlocks were stunted abominations, deformed and morally depraved. The novel—in which cannibalistic and bestial Morlocks prey upon the vegetarian Eloi aesthetes—has been read as a "blue-print" of degenerationist concerns and a critique of "outcast London."[47] It is also a parable about the loss of energy and devitalization. Wells frequently examined the impermanence, imperfection, and insignificance of human life. A great admirer of Darwin and a one-time student of his "bulldog," T. H. Huxley, Wells read into Darwin a legitimation of ephemerality, degradation, and teratology. Set on an island near the Galapagos,

Wells's *The Island of Dr. Moreau* (1896) is inhabited by monstrous human-animal hybrids. The "beast people" include Leopard Man, Dog Man, Puma Woman, Monkey Man, Wolf Woman, and the "Mare-Rhinoceros-Person," a world "teeming with abominations."[48] Wells, like Lombroso (whose theory on female insensitivity Wells had written about), theorized "a human body both chaotic and entropic, both hybridized and prone to reversion."[49] The novel's protagonist Prendrick devolves during the narrative, becoming hysterical, nervous, and ineffectual, hairy and agile, and accustomed to sleeping in a den. Moreau—the novel's criminal genius—was the personification of nature. He represented the apex of human intellectual evolution but was as "inhuman" as his grotesque creations, bereft of any of the civilizing human emotions of compassion.[50] Wells explained that the novel was "written just to give the utmost possible vividness to that conception of men as hewn and confused and tormented beasts."[51] In the end, the novel accomplishes the ruination of the human subject, "without apology, without nostalgia, without remorse."[52]

Like psychoanalysis, that fin-de-siècle science of the unconscious mind, what characterized the Gothic was a confessional mode, the impulse to unburden oneself of unpalatable truths. Revolving around the insecurities of biomedical knowledge and often featuring young, inexperienced medical doctors, Gothic novels featured an abundance of primary texts: letters, journals, interviews, official reports. It is as though the prolix archive of established medicine has had to be reopened; epistemology had to follow a cadaverous ontology and return itself to an earlier, primitive state of contestation. Once the body of knowledge has been exhumed, its disagreeable corporeality becomes evident. Horrific truths about human nature provide neither sensational thrills nor spectacular forewarnings but rather affect a nauseating collapse of meaning.[53] The war against crime was allied to the administration of abnormality, expanding social control as it relaxed moral judgment.[54] In this context, Goring's *The English Convict* can be read less as criticism of Lombrosoian science than as its rhizomatic outgrowth. For if criminal anthropology had blurred the boundary between the normal and the abnormal, then biometrics simply located these categories on the same normal distribution curve, the one imperceptibly blending into the other.[55]

In 1892, in a measured article warning of the dangers of over interpreting criminal statistics, the Rev. W. D. Morrison expressed a skeptical note about Lombroso's theory of the born criminal.[56] "Whatever may be the ultimate fate of Lombroso's theory," wrote Morrison, "he has unquestionably succeeded in calling attention to the fact that a larger proportion of anomalies is to be found among the criminal population than among ordinary members of the community." A "debilitated body," after all, had "a tendency to produce a perverted mind." But instead of regarding criminal stigmata as signs of degeneration, Morrison proffered a sociological version of events, suggesting that such marks might predispose a person to a life of crime simply as a result of social prejudice, a dearth of employment opportunities, and an embittered sensibility: "In the inevitable and unceasing struggle for existence a considerable proportion of the feeble, the degenerate, the malformed, the anomalous are not fitted for one reason or another to earn a living by normal methods, and society looks upon all who adopt abnormal methods as criminals." Physical anomalies among offenders were neither evidence of their mental capacities nor did they support for the existence of a criminal type. Physical abnormalities were "proof of a fact apparent everywhere, that the physically anomalous and incapable are less adapted to fight the battle of life, and are accordingly more likely to come into collision with the law."[57] Visible stigmata were the cause of a criminality acquired through the hostile actions of the prejudiced.

Morrison argued that although it was widely believed that crime had fallen over the previous thirty years, this apparent decline could not be taken for granted without taking into account changes in judicial procedure and methods of incarceration. Changes in the law had criminalized some activities, while some criminal offences had been rendered harmless. In 1890, for example, proceedings against over eighty thousand parents had been initiated for not sending their children to school, a statutory requirement since the passing of the 1870 Elementary Education Act. An increase in the number of sexual crimes had occasioned the Criminal Law Amendment Act of 1885, but the police were increasingly turning a blind eye to intemperance. The use of reformatories and juvenile homes challenged the usefulness of the incarceration statistics as a guide to crime rates. Despite living in an age when statistics were collated on a vast number of criminal subjects, those statistics could never be taken at face value. The more crime was studied the more obscure it appeared.

In 1893, *Science* carried a short review of Arthur MacDonald's *Criminology*. In his brief introduction to the book Lombroso had defended his theory of the criminal type, "the organicity of crime, its anatomical nature, and degenerative source." The anonymous reviewer, however, claimed that the notion had been "distinctly rejected" by the criminal anthropologists assembled at the previous summer's Brussels congress, and it was "encouraging to note" that MacDonald now considered the criminal type "from the psychological rather than the physical side.... This is virtually giving up the position of Lombroso, which, in fact, is no longer defensible. There is absolutely no fixed correlation between anatomical structure and crime, so far as has yet been shown."[58] Bemoaning the lack of progress in criminal anthropology a few years later, Gabriel Tarde concluded that "Lombroso's alleged criminal type is a chimera; that the Italian school is engaged in the desperate undertaking of rescuing from perdition an error which it knows to be an error, but which it hopes, in spite of the head of the school, to attach to some theory palpably less visionary. This can not be done."[59] The novelists had critiqued *homo criminalis*: now it was the turn of the criminologists.

"Prominent among the new 'sciences' that have sprung up, mushroom-like, in our generation is 'criminology.'" Thus began Dr. H. S. Williams' 1896 *North American Review* article, "Can the Criminal be Reclaimed?"[60] "Its advocates regard the criminal as a distinct type of the *genus homo*. With the true spirit of our induction-haunted age, they analyze the delinquent out of all association with his fellows, making him stand apart as a separate order of being." Williams ridiculed the claim that criminologists could pick out the murderer before he had committed a murder or the inebriate who had never been intoxicated, adding: "No doubt the abstract robber, forger, and what not, will follow in due season." "All this is very delightful if true. It suggests visions of a golden age, when science shall rule society and by its anticipatory action nip all crimes in the bud by restraining their would-be perpetrators. But with the vision comes also a doubt."[61] Williams suggested that crime was not a particularly special phenomenon, "but merely the expression of a relation." Criminality was a matter of social consensus: "deeds stamped as criminal under some circumstances are justified under others." Killing in warfare, for example, or accidental homicide, were not considered crimes. Furthermore, since all human actions involved advancement for personal gain, a criminal act could not be singled out as particularly special: "sinfulness is fixed in accordance with an absolute standard of ethics, criminality in accordance with a relative or human standard."[62] Because ethical codes were constantly changing, the boundaries of criminal action were forever fluctuating. It was therefore impossible to define crime accurately. Williams accepted that it was a fact of human psychology that people had criminal desires, but these were repressed by the pressures of social life. "The spirit of practical altruism was born, and the cornerstone of civilization was laid."[63] Persons who had no moral sense were known as habitual criminals. According to criminology,

Williams wrote, people became criminals because of "inherently defective" brains. He proposed "that in the great majority of cases they have failed to evolve because they were human and could not rise far above their environment."[64] There was good and bad in everybody, but the two extremes of the social scale were not morally distinguishable: "Wax to receive, marble to retain, that young mind was graven deep with the lines of wrong living. A subtle poison permeated every cell of its body; what wonder if it thenceforth gave out none but poisoned thoughts?"[65] Careful training was necessary if the best results were to be obtained because the "flowers of the human mind do not bloom on human weeds."[66] Williams pointed to the forced feeding of a larva in the hive that produces the queen bee, drawing a parallel with the process of socialization early in infancy and early childhood. It was "familiar knowledge that most wild beasts can be tamed only if taken while young," "The day is past when it was supposed that the human mind is intrinsically different in kind from other animals. It is now known that general biological truths apply to each and every member of the organic scale from highest to lowest.... And in this connection it may not be amiss to note that the human family contains but a single species."[67]

Practical humanitarians had rescued thousands of "vicious little John Does" from vice and developed them into useful citizens. To guard against mental disease by educating the mind and by avoiding pursuits that bring intellectual strain was a "problem of practical sociology." "If this view is correct, the criminal differs from his fellows not so much in inherent depraved tendencies as in defective powers of resistance. The law-abiding individual has or has had many of the same propensities that are patent in the criminal, but they have been corrected, repressed, or even eradicated by the cultivation of higher instincts; that is, by ethical development."[68] A considerable degree of development was always possible, even in habitual criminals. A child reared by criminal parents would become a criminal, but there was no need to invoke heredity in explaining such cases. Statistics were at hand to prove that "even after the plastic period of childhood has been spent in the haunts of vice ... much may still be done in many cases to develop the higher ethical sense upon which depends the resistant power that shall shield from crime." Williams pointed to the records of the Reformatory at Elmira, New York. Of all the persons admitted, the records showed that forty percent had no moral sense; many had an ancestry made up of epileptics, the insane, or drunks. Of these "moral imbeciles," sixty-seven percent were illiterate or could read and write only with difficulty. After treatment, the authors claimed, eighty percent of them were returned to the world, and of these, four-fifths become permanent honest breadwinners. "When each penal institution in the land has come to be such an ethical factor as this," Williams enthused, "the records of criminology will tell a very different story from the doleful one they now present."[69]

In 1898, Frances Kellor of the University of Chicago expressed similarly skeptical views about the differential rate of criminality among women compared to men, dismissing those anthropologists who considered woman to be "midway between child and man in her development."[70] The information on female criminality that had been gathered up to that point was inadequate, because there were "numerous conditions and acts indicating a low morality and a criminal nature, acts which are injurious to the community and state, of which the law takes no cognizance, but which would be indispensable factors in an accurate study of comparative crime in the sexes." Once such factors had been taken into account, the variation in the statistical rate of crime for the two sexes would be eliminated: "Were the determination of the criminal and non-criminal classes less legal and more sociologic; less based upon the crime and more upon the criminal, and upon the amount and nature of crime rather than upon the number of convictions, which are less than one-third of the actual number of crimes committed, the proportionate rate of the crime in the two sexes would be much more uniform."[71] Dugdale's study of the Jukes family had recently demonstrated that criminal tendencies were inherited equally by the two sexes.[72] To a considerable extent, the apparent difference in criminal statistics was an artifact of the "unequal political privileges" enjoyed by men. Because of their social standing, women simply did not have the opportunity to commit offences against the government, "including violations of the election and postal laws; of the revenue laws; against public health, as adulteration of foods." As they entered vocations hitherto open only to men, their rates of offending went up.[73]

In 1900, Kellor published the results of a meticulous investigation, "Psychological and Environmental Study of Women Criminals."[74] The assiduous researcher had interviewed inmates at five institutions: the reform school at Geneva, the penitentiary at Joliet, the workhouse at Cincinnati, the Ohio State Penitentiary at Columbus, and the workhouse and penitentiary at Blackwell's Island, New York City. She had also performed numerous psychological tests on her subjects: sixty-one female offenders and fifty-five students for comparison. Most previous investigations among criminals had been "upon the anatomy," Kellor explained, to the neglect of "the functioning in society, the mental, moral, and emotional nature." As a result, the criminal had come to be regarded as "a finished product" rather than an individual "in a state of evolution."[75] The researches of the European criminologists such as Lombroso, Ferri, and Tarnowsky had been largely accepted, but their findings were questionable Kellor maintained, because this work had ignored "the mental and emotional impulses [and] the tremendous forces of social and economic environment."

Kellor implied that the data collected by criminal anthropology was inaccurate, and that it promoted an illiberal and overly theoretical agenda. What was required was a "criminal sociology": a sociology informed by psychology. "Psychology makes possible a quantitative sociology." Such a project would have to study the noncriminal class, something that criminal anthropology had failed to do.[76] According to Kellor, one advantage of psychological testing was that it led inevitably toward explanations in terms of social factors. The ascertainment of a condition "such as defective hearing or taste," for example, "revealed methods of living, habits, disease, etc." The extensive anthropometric measurements Kellor took included "weight, height, sitting height, strength of chest, hand grasp, cephalic index, distance between arches, between orbits, corners of eyes, crown to chin, nasal index, length of ears, of hands, of middle fingers, of thumbs, width and thickness of mouth, height of forehead, anterior and posterior diameters."[77] The measurements sought a knowledge of the five senses and the capacity for perception, coordination, and adjustment. Because anatomical differences between people "depended upon race, climate, soil [and] nutrition," it was important to make comparisons within racial groups, not between them. To Kellor, it was evident "that we cannot accept the statements that criminals are more brachycephalic than normals, when one has measured only Italians or Russians, and that this is an ethnic characteristic. It is just here that Lombroso's results are untrustworthy when applied to various races and countries."[78]

Kellor disputed Lombroso's proposals that prostitutes were heavier, longer-lived, and more likely to be left-handed than other women. She criticized his methods of data collection, presented data in opposition to his, and questioned his interpretations: "He asserts that prostitutes possess, more frequently than normals, enormous lower jaws, projecting cheekbones, projecting ears, virile and Mongolian physiognomy, prehensile feet, masculine voices and handwriting. I am unable to verify these, and think that, especially in the first-named, racial influences again operate. I found faces with hard expressions, and voices harsh and cynical, but they did not possess the peculiar

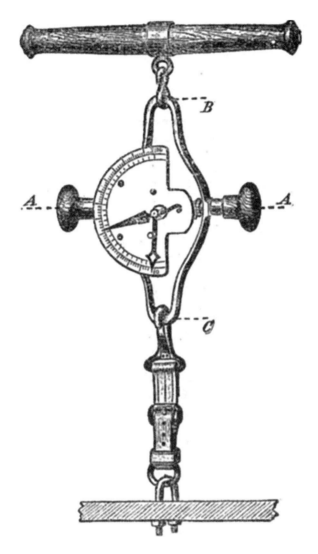

Arthur MacDonald's back, chest, and leg dynamometer, a device for measuring muscular power (1898). From Arthur MacDonald, "Psycho-physical and Anthropometrical Instruments of Precision in the Laboratory of the Bureau of Education" (1898), chap. 5, p. 1186, fig. 78. In: *The Experimental Study of Children*, 1141–1204 (United States of America, Bureau of Education, Report of the Commissioner of Education, 1897–98).

masculine quality, and I do not believe that harshness, cynicism, coarseness make them masculine."[79] In opposition to Mantegazza's claim that women criminals were "almost always homely, if not repulsive," Kellor asserted that "the effect of clothing, cleanliness, etc., must be considered, as deficiency in these particulars renders even a normal individual unattractive."[80] Having secured her subjects' confidence and consent—"for they were extremely suspicious and superstitious of any investigation, especially where instruments were used"—Kellor set about measuring their psychological capacities. Her psychological tests included those for memory, color blindness and color preference, sensibility of the skin, the five senses, fatigue, pain, respiration, and the association of ideas. Kellor suggested that her study was "the first attempt to secure a series of such measurements from female delinquents, and to compare the results with those from a different educational and social stratum of society."[81]

Differences in memory were explained by noting the differential effects of education, the difference in color preferences to economic and social factors. The qualities of women she discreetly referred to as "courtesans" she ascribed to what was "current coin within their community." Although Kellor confirmed Lombroso's finding that criminals had greater insensibility compared to normal women, she attributed the difference to the nature of the occupation, poor diet, and inadequate care of the body, all of which were bad habits that tended "to render the sensibilities less acute." She also pointed out that it was difficult to maintain her subjects' concentration on this particular task—poor concentration tended to render the results inaccurate—and she found that flabby skin confounded the results. The tests on taste and smell resulted in greater differences between delinquents and students than in any of the other tests: "Instead of proving one of the current theories, that the criminal is allied to the savage, and is more dependent upon physical senses than upon his intellect, and thus has these more acutely developed, I found them to disprove it. In taste the delinquents were only about two-thirds as accurate as the students, in smell only about one-half."[82] Kellor put the difference down to the use of snuff, alcohol, and tobacco, "coarse, strong foods," poor sanitary conditions, "unsavory odors," and disease. She patiently offered an alternative sociological explanation for every criminal anthropological claim that she examined. For example, the assertion that "when women are criminal they are more degraded and more abandoned than men" could be attributed "to the difference in the standards which we set for the two sexes." "We say woman is worse, but we judge her so by comparison with the ideal of woman, not with a common ideal. For instance, I have included swearing and use of tobacco as bad habits among women; among men we should not consider them in the same light. These make a deeper impression by reason of the requirements of *our* ideal, not in the light of plain fact."[83]

From a social point of view, an intoxicated woman merely appeared more degraded than a man in the same state; she wasn't inherently so. Because standards of licentiousness, manners, and cleanliness varied across the social classes, it was illegitimate to regard women in general as more degraded than men. Criminal anthropology had also ignored the economic conditions that rendered women more "liable to immorality." Kellor gave a stark illustration of one such pressure that gentle readers of academic periodicals might not have been aware of: "Clipping from the newspaper some thirty advertisements for clerks, stenographers, bookkeepers, etc., I had an assistant answer them. Although she visited only a little more than half the places, almost every one of them was a snare for immoral purposes, and the proposals were so bluntly made that she declined answering more."[84] It was difficult for a woman seeking economic independence to resist considerable pressures from numerous quarters.

At the heart of Kellor's critique of criminal anthropology lay a conviction—by then shared by many novelists and scientists alike—that it was no longer legitimate to conceive of the criminal as a separate sort of human being. She had failed to replicate almost all of criminal anthropology's claims in this respect. But there was one final test in Kellor's psychological and anthropometric arsenal that pointed toward a new way of thinking about the relationship between the human body and crime: "The last, and perhaps most interesting, test was that made with the kymograph, an instrument designed to register the respiration curve upon smoked paper." The instrument consisted of a clockwork base and a drum, around which was a roll of smoked paper. Resting lightly against the smoked paper was a pointer connected by rubber tubing to a respirator fastened onto the subject's chest: "When the subject inspires and expires, the air is forced down and back the tube, the pointer making a curved line upon the paper as the drum revolves. Every change in the amplitude and the rate of the breathing is thus graphically portrayed." The test was given to determine the amount of emotional reactions to stimuli, as shown by changes in breathing.[85]

Five breathing curves were recorded for each subject. Considering the humanitarian impetus of Kellor's investigations, it was surprising that this final experiment involved inflicting emotional and physical pain on her participants. For the first curve, the subjects were asked to sit quietly and think of anything except the experiment. They were asked to explain what they were thinking about when the curve exhibited a marked depression or rise. Toward the end of this phase, "suddenly a block was dropped, or a hammer struck," and the resultant effect was noted. The second phase was designed to produce a normal curve, the subject being asked to read a newspaper clipping designed "not to excite undue interest." Toward the end of this stage, the subject experienced an unexpected "sharp pinprick in the back of the neck." The third curve investigated "the difference between the various methods of reading," while the fourth "was related to the effect of interest." Here the subject was required to read a dull agricultural newspaper report followed by a graphic account of a prison rebellion. In the final part of the test, Kellor furnished her subjects with two ideas to think about. The first suggestion—concerning the joy of imminent release—almost always produced "deeper and wider" curves. The second suggestion reminded the inmate of her shame, "the fact of her being in such an institution and its effect upon relatives and friends." At least three of her subjects wept at this point.[86] Finally, to induce fear in her subjects, Kellor deployed what she called "a simple device": "Placing a plethysmograph near the temple, I said I intended applying an electric current; that if they would not move or speak the pain would be slight. The fear of electricity is very great, and this never failed. My great difficulty was in keeping them quiet, so excessive was this fear. Two marked changes were observable: either the curve became almost a straight line, as when they held their breath, or it became ragged and of varying amplitude, as when they became nervous through fear."[87]

Kellor had discovered that it was sometimes difficult to obtain truthful biographical information from her subjects. It was generally "difficult to secure a definite statement" because "the women generally desire to convey an impression of their superior intelligence."[88] The inmates' verbal assurances could not be taken at face value. Almost by accident, however, and only as a secondary fact-gathering exercise, Kellor had discovered that her various tests could uncover hitherto hidden facts: "From such tests as those for memory, association of ideas, reading, respiration, etc., I was able to judge if their statements were true."[89] This innocent but portentous statement is buried in a mass of experimental detail. It is the first example of a physiological test applied to the examination of verbal statements within a criminological discourse that rejects the idea that the criminal is a special type of person.

Kellor's research was undertaken at the end point of a literary, practical, and empirical movement that challenged the idea that there was a separate species of person predisposed to crime.[90] Novelists had suggested that everyone was potentially capable of criminality. Sociologists had argued that social, economic, and political factors influenced crime rates and criminality. Kellor would go on to have a widely celebrated career as a political activist and an immigration and arbitration specialist.[91] But her work at the end of the nineteenth century, by accident, had posited the idea that physiological tests could be used to uncover secrets and lies within the minds of ordinary people. Frances Kellor had made the lie detector possible.

CHAPTER 5

"To Classify and Analyze Emotional Persons"

The Mistake of the Machines

~~~~~

"I've been reading," said Flambeau, "of this new psychometric method they talk about so much, especially in America. You know what I mean; they put a pulsometer on a man's wrist and judge by how his heart goes at the pronunciation of certain words. What do you think of it?"

"I think it very interesting," replied Father Brown; "it reminds me of that interesting idea in the Dark Ages that blood would flow from a corpse if the murderer touched it."

"Do you really mean," demanded his friend, "that you think the two methods equally valuable?"

"I think them equally valueless," replied Brown. "Blood flows, fast or slow, in dead folk or living, for so many more million reasons than we can ever know."

—G. K. Chesterton, *The Mistake of the Machine* (1914)

By the time G. K. Chesterton wrote his Father Brown detective story, "The Mistake of the Machines," an instrument known as an "electric psychometer" had acquired a notorious reputation in the American press. Newspaper reporters were fascinated by this and other "soul machines," "truth-compelling machines," and "machines to cure liars."[1] But none of these devices were lie detectors in the sense that the term has been understood since the 1920s. There were many continuities between the years these instruments attained cultural prominence (1900–20), and the subsequent period when the lie detector achieved widespread fame (1920–50). But there was also one crucial discontinuity: the lie detector's advocates would disown the notion that still preoccupied the scientists described in this chapter, namely, the born criminal.

The first use of the term "lie detector" appears to have been in Charles Walk's novel *The Yellow Circle* (1909).[2] As early as 1910, some of the machines described below were being used to detect lies—but not by scientists. It was the authors of pulp magazine stories who explored this new possibility. Before about 1915, criminal science had yet to completely abandon its belief in *homo criminalis*. The scientists described in this chapter used instruments in the hope that they would reveal deviant pathologies of character. Fiction writers, however, were already describing how the very same instruments could be used to detect guilty secrets from the minds of otherwise normal wrongdoers. This fundamental innovation was made in mass circulation magazines. This under-appreciated art form was entirely at ease with the notion that a criminal was merely someone who had been caught out, either in the act of committing a crime or while remembering it later. The revelation that criminals were normal people remained too disagreeable a proposition for positivist science until after the First World War.

Early in June 1907, a reporter for the Sunday *New York Times* paid a visit to "the famous alienist," Dr. McLane Hamilton.[3] He had come to investigate a "mysterious little machine" that was beginning to attract "the attention of the experts in mental diseases." The reporter was somewhat intimidated by the "electric psychometer," he admitted, despite it being "a simple, innocent looking affair." "I don't want the thing tested on me," he proclaimed, "as he sank, rather nervously into one of the capacious chairs in Dr. Allan McLane Hamilton's studio." He then explained the reasons for his unease with the "electric measurer of the soul": "An infallible detector of crime and insanity is ever an uncomfortable thing to come into contact with, and there was a sort of suggestion of inexorable fate in the various scientific appliances and the enigmatic books that furnished the room that made one almost wish for an era when such things as electric psychometers were unknown."[4] Despite his misgivings, the apprehensive reporter contributed to the growing excitement about the new technologies for analyzing the criminal mind and diagnosing mental disease. "It is barely possible," his piece began, "that Judges and all the paraphernalia of the criminal courts will become superfluous before the century is over," their place taken by "this new invention, 'The Electric Psychometer,' whose inexorable 'finger of light' has already revealed more than one culprit to his baffled accusers."[5] "There are few branches of science that have been developed more radically or more quickly than psychology," asserted Dr. McLane Hamilton. "The old method of examining into morbid conditions was crude." An "entire revolution" was held to have brought to the study of the mind "an almost mathematical certainty." "There is hardly a mental function that cannot be determined by a special apparatus, so that now we are able not only to detect the acuteness of perception and the activity of the senses in individual cases, but to actually gauge the rapidity of thought and the specific association of ideas."[6]

Another type of "special apparatus" had recently been developed by a young Swiss physician. The aim of Carl Jung's investigations was to discover "emotional complexes" in his patients. While his colleague Sigmund Freud had used word association purely as an aid in the diagnosis of unconscious neurotic conditions, Jung's "ingeniously contrived suggestions" succeeded "in awakening sub-conscious trains of thought which in some cases ... caused the patient to make declarations of secret matters, and even confessions of guilt."[7] It had proved its efficiency "in the hands of Jung, its discoverer," prompting him to "assert his conviction that the system is adequate when employed in the detection of crime." But without the aid of apparatus, the word association method lacked accuracy and precision. The response time to the stimulus word indicated "the relative quickness of that patient's intelligence" and the response word itself "the association of ideas which had been aroused," but there was no method of measuring "the degree or quality of emotion caused by the test word."[8] Jung's solution was to use the word association method in conjunction with "an instrument of the utmost delicacy," whose operations had already set the scientific world "agog with curiosity": the galvanometer.

By the time Jung was engaged in this research, galvanic phenomena had been observed in clinical patients for over twenty-five years. Romain Vigoroux, a pupil of Duchenne de Boulogne at the Salpêtrière Hospital, apparently made the first observations of psychological factors in relation to electrodermal phenomena in 1879.[9] The technique involved using two electrodes to apply an electrical current of about two volts to the body while measuring concomitant electrical changes with a galvanometer. In 1880, while investigating changes of excitability of motor and sensory nerves, A. D. Waller—who would later become the first physiologist to make an electrocardiogram—had observed irregular galvanometric deflections, which he attributed to alterations of contact between the skin and the electrodes.[10] In 1888, Charles Féré, a pupil of Charcot, suggested that the deflections were caused by a lowering of the resistance of the body under the influence of emotional states such as those found in hysteria.[11] Having applied no external current, in 1890 Jean de (Ivan) Tarchanoff claimed that the phenomenon occurred in all normal persons. In 1897, G. Sticker attributed the perturbations of the galvanic current to the actions of the capillary system of blood vessels.[12] "Whoever is from any cause emotionally roused on looking at a picture will react with a definite increase of the current," he wrote, "whilst whoever is unmoved by the picture, or in whom it arouses no memory, will have no skin excitation."[13] R. Sommer concluded in 1902 that no psychic influence on the phenomenon could be established with certainty.[14] Féré's work was rediscovered in 1904 by E. K. Müller, and Veraguth labelled the phenomenon the "galvano-psychophysical reflex" in 1906, the year in which Jung began his investigations.[15] Jung's association experiments inspired a great deal of work in the field because it suggested that aspects of "mental" life could be revealed by electrodermal responses.

Jung attached the galvanometer to his patient while applying the word association method.[16] This combination of hard and soft technologies resulted in the hope that "the suspected criminal or the mental pervert who is the subject of investigation" would be betrayed by electric forces so subtle as to be discernible only through the application of a scientific instrument: "Ordinarily invisible, it is this inner, uncontrollable manifestation which impresses itself upon the electric current that is passing through him, the effect of which is quickly shown, to those who are watching the experiment, in the sudden agitation of the moving ray of light and in the irregular, jagged line—the record, it may be, of the man's guilt—which is rapidly traced on the revolving cylinder."[17] Some words produced no effect upon the "finger of light" of the psychometer. Others, however, struck "some emotional complex deep in the soul of the individual experimented upon" and sent "the light along the scale for a distance of one centimeter up to six, or eight, or more, in proportion to the intensity and actuality of the emotion aroused." Under this arrangement the boundary between "the suspected criminal" and "the mental pervert" was quite indistinct. Working in the tradition of the clinical experiment,[18] Jung used physiological instrumentation to investigate what he called the "emotional complex."[19]

"It is almost like sorcery," said Professor Frederick Peterson of Columbia University, whose own laboratory was awaiting the delivery of Jung's "marvelous machine."[20] His allusion to the magical properties of the new technology was echoed by the newspaper reporter who had thought the instrument's operations reminiscent of "the uncanny movements of a planchette board, whose writing some fifty years ago was thought to be the direct result of a conspiracy on the part of the Evil One." Trying to make contact with the dead using the Ouija board, however, was but a crude game compared to what could be accomplished with the new machine. "Although the electric psychometer is a new invention," said Peterson, "it has fortunately gone through a sufficient number of tests to leave no doubt as to its genuine value. It positively does what its discoverer claims it can do, and there thus appears to be no limit to the possible uses to which it may be put in the future wherever the detection of crime or the treatment of mental and nervous diseases engages the attention of mankind." For Peterson, as for Jung, the detection of crime and the treatment of mental diseases were one of a kind: there was no essential difference between the two projects. At the center of his inquiries Peterson situated specific types of people, as his 1908 article in the *British Medical Journal* demonstrated. He believed the apparatus was far more accurate for recording physical changes associated with mental function than any already employed, such as the plethysmograph and the pneumograph.[21] "Furthermore, its distinctive field seems to be that of recording the effects of emotions upon the organism. . . . I have given the name 'electric psychometer' to the galvanometer thus used." How such methods might prove useful in attaining knowledge "of hidden matters in the minds of neurasthenics, hysterics, and criminals," was impossible to foresee; but advocates held that this new and valuable method of exploration in psychology was already "beyond question."[22]

Peterson also collaborated with the Wundt-trained Yale psychologist E. W. Scripture, whose enthusiasm for the new psychology of the 1890s had "contained an element of exaggeration and another of egotism," according to historian of psychology E. G. Boring.[23] Scripture's two popular books, *Thinking, Feeling, Doing* (1895) and *The New Psychology* (1897), were both "packed full of pictures of apparatus, graphs and other apt illustrations," but evidently contained "no argument, no theory, no involved discussion."[24] Scripture's 1908 paper, "Detection of the Emotions by the Galvanometer," posited three goals: first, the construction of a self-registering instrument "as complete and handy as a chronoscope"; second, to "definitely settle" the issue of which bodily changes produced which emotional effects; finally, to "classify and analyze emotional persons."[25] "Medical men have so much material with diseased and abnormal minds that we can work most profitably on the pathologic side," he maintained.

In one of his first papers on the "psychophysical relations of the associative experiment," Jung described Veraguth's investigations of the "galvano-psychophysical reflexes."[26] In the course of these experiments, Jung reported, "it was discovered that the action of the galvanometer was not in direct relation with the strength of the irritation, but more especially with the intensity of the resulting psychical feeling tone."[27] Jung was particularly interested in this "psychical feeling tone." His aim was to use the galvanometer to diagnose pathological "complexes of ideas," ultimately in the hope of treating *dementia praecox*—the condition that Eugen Bleuler would rename "schizophrenia" in 1908. The reaction time association experiment, Jung wrote, "is a good means of fathoming and of analyzing the personality.[28] According to the opinion of some German authors, this method should be applied for the purpose of tracing the complexes of culpability in criminals who do not confess."[29] Jung's main concern was evidently the diagnosis of mental pathology, not the unmasking of criminals per se.

In his next paper, a collaboration with Peterson, Jung hoped to compare the galvanometric and pneumographic curves, simultaneously recorded under the influence of various stimuli. The goal was a comparison between normal and insane individuals.[30] Jung's experimental group consisted of patients suffering from epilepsy, dementia praecox, general paralysis, chronic alcoholism, and senile dementia.[31] "The galvanic reaction depends on the attention to the stimulus," he concluded, "and the ability to associate it with other previous occurrences. This association may be conscious but is usually subconscious."[32] But the researchers found the work frustrating due to "nervous tension," the "forced and artificial situation of the test," and other "inexplicable influences at work." Nor were they impressed with their abilities to decipher the relationship "between the galvanometric and pneumographic curves." A "surprising divergence between the influences at work upon them" produced inconsistencies. With resignation, they reported having studied hundreds of different curves, carefully measuring the length and duration of each inspiration, without managing to decipher any regular relationship or reliable correspondence between them.[33] Their aim was not to ascertain guilt but rather to document "the emotions of the subconscious, roused up by questions or words that strike into the buried complexes of the soul."[34]

Less than three months after the *New York Times* had reported on Jung's experiments, the newspaper carried a sensational article headlined "Invents Machines for 'Cure of Liars'/Prof. Munsterberg Experiments to Reduce Knowledge of Truth to Exact Science/Way to Detect Criminals."[35] "Hugo Munsterberg of Harvard University has just crowned the achievements of a life devoted to psychological research by the invention of several little machines to record the emotions and reveal the secrets of the human mind." One scientist called them "Truth-compelling machines," another "Machines to cure liars." Because it was held that Münsterberg had reduced a knowledge of the truth "to an exact science," claims were made that in a few years "no innocent person will be kept in jail, nor, on the other hand, will any guilty person cheat the demands of justice."[36] Münsterberg was not averse to making similarly sensational statements himself. "To deny that the experimental psychologist has possibilities of determining the truth-telling powers" he told the *Times* reporter, "is as absurd as to deny that the chemical expert can find out whether there is arsenic in the stomach." In comparing the experimental psychologist to the chemical expert, Münsterberg was placing psychological knowledge on a par with chemistry. But the existence of so many apparently different "truth-compelling machines" suggested that "the truth" could be obtained through a variety of means. Designed to record the "involuntary writings of the suspect," the automatograph, for example, consisted of a wooden sling suspended from the ceiling upon which rested the suspect's arm. The "pneumograph" recorded variations in breathing caused by emotional suggestions, while the "sphygmagraph" recorded the action of the heart. The last instrument, probably a galvanometer, "employ[ed] electricity in the form of an instrument somewhat resembling the recording telephone."

Münsterberg's work was sufficiently well known for it to be satirized in the press. A reporter from the *New York Times* described a visit he had apparently made to "Prof. Hugo Monsterwork" at his home to see his "machine for distinguishing true statements from false." Monsterwork enthused about his "great invention . . . the wonder of the age" that he had discussed at a recent meeting of the "German-American Metaphysical, Dabbledabble and Subsequent Beer Society." Some of the other marvels on display at "Villa Psyche" were "liar-detecting machines" such as the "Fibbograph" and the "Mendaxophone," and other more specialized devices such as the "Ancestrophone," the "Courtesycometer," and the "Pie-crustograph."[37] Münsterberg responded to the criticism by accusing the newspapers of having a deluded fascination with invention: "All these instruments of registration have belonged for decades to the ordinary equipment of every psychological laboratory," he explained. "It was therefore a sad commentary when, recently, scores of American papers told their readers that I invented the sphygmograph, the automatograph, and the plethysmograph last summer—they might just as well have added that I invented the telegraph last spring. To recent years belongs only the application of these instruments to the study of feelings and emotions."[38]

Münsterberg's interpretations of the various tests demonstrated that he was committed to a personological theory of criminal types: "The emotional retardation of suspicious associations, characteristic of the average criminal, was, as expected, entirely lacking in this wholesale murderer. That does not mean that he lacks feeling; my experiments showed the opposite. To be sure, his sensitiveness for pain was, as with most criminals, much below the average."[39] Münsterberg believed in "average criminals" who had specific mental and physical characteristics, evidenced in his commitment to a Lombrosoian discourse of criminal difference. Münsterberg's machines were not lie detectors. They were "machines for the cure of liars," "truth-compelling machines," or even "scientific crime detectors" according to *Scientific American*,[40] but they did not detect lies in the normal person. Münsterberg had even said, "The word lie is not in my lexicon."[41] He was defending himself against accusations of mendacity, but the target of Münsterberg's researches was not the lie, but the personological notion of the morbid "complex of ideas" as indicative of a deeper psychopathology.

Criticized by his colleagues for writing what they dismissed as "yellow psychology"—popular psychology written for magazines and newspapers— Münsterberg felt obliged to explain the new science to the general public in mass circulation periodicals.[42] The German émigré had eclectic psychological interests, and during 1907 and 1908, he published a number of articles in the popular press on the relationship between psychology and crime.[43] His first such piece, "Nothing But the Truth," discussed the bearing of psychological perception experiments on the reliability of witness testimony.[44] The article inspired a *New York Times* commentary headlined "A Psychologist's Judicial Warning / Munsterberg Writes of Errors in Observation Due to Personal Equation."[45] "In short," Münsterberg concluded, "every chapter and sub-chapter of sense psychology may help to clear up the chaos and the confusion which prevail in the observation of witnesses."[46] So divergent were the performances of subjects on simple tasks of visual, aural, and temporal perception, that the truth of perception was essentially a matter of individual differences. Münsterberg wanted experimental psychology to spearhead a social revolution that would bring about efficiency, justice, and progress.

Psychology was considered progressive because it produced an objective and reliable knowledge of human nature, and also because it was humane. Münsterberg's next article, for *McClure's*, titled "The Third Degree"[47] argued that the "clean conscience of a modern nation rejects every . . . brutal scheme in the search of truth"; here the Harvard psychologist was twenty years ahead of his time. Appeals against the brutalities of the third degree—the practice of beating confessions out of suspects—would not become commonplace in the popular press until the late 1920s. But as Münsterberg put it, objections against violent methods of obtaining confessions were not based on "sentimental horror" or "esthetic disgust," but "the instinctive conviction that the method is ineffective in bringing about the real truth."[48] What was effective were "the methods of measurement of association which experimental psychology [had] developed in recent years."[49] Invoking medical metaphors, Münsterberg predicted that the chronoscope "will become more and more, for the student of crime what the microscope is for the student of disease. It makes visible that which remains otherwise invisible, and shows minute facts which allow a clear diagnosis. The physician needs his magnifier to find out whether there are tubercles in the sputum: the legal psychologist may in the future use his mental microscope to make sure whether there are lies in the mind of the suspect."[50]

Theoretically capable of measuring to a thousandth of a second—though in practice this level of accuracy was extremely difficult to achieve[51]—the chronoscope was used to measure the length of time it took a subject to utter an associated word in response to a stimulus word. Through the "exact and subtle study of mental associations . . . a deep insight" could be "won into the whole mental mechanism."[52] The content of the associations was as important as the time it took to produce them: "Those words which by their connection with the crime stir up deep emotional complexes of ideas will throw ever new associations into consciousness, while the indifferent ones will link themselves in a superficial way without change."[53] In such a way "the mind betrays its own secrets." Münsterberg illustrated his methods with a description of how he obtained a confession of guilt from Harry Orchard, the man who had confessed to the 1905 murder of the ex-governor of Idaho, Frank Steunenberg. Millions of readers across the country were following the trial.[54] "I began with some simple psychological tricks," Münsterberg wrote, "with which every student of psychology is familiar." Having shown the murderer some "tactical illusions," the psychologist claimed he had managed to bring the man "entirely under the spell of the belief that I had some special scientific powers."[55] "The time will come," he concluded, "when the methods of experimental psychology cannot be excluded from the court of law."[56]

Hubristic predictions about psychology's impact on the legal system were regularly made in the press. A 1907 *New York Times* article, "Applied Psychology and Its Possibilities," was typical.[57] "Recent Discoveries in Mental Science

Lay Bare the Mind of the Criminal to the Psychic Expert" proclaimed the headline, "Remarkable Part Which the Modern Laboratory May Play in the Great Court Trials of the Future." "Psychology is most important in its application to legal evidence," said Professor R. S. Woodworth of Columbia University. "There are a number of difficulties connected with the testimony of witnesses which are essentially psychological difficulties."[58] Psychologists had lately been turning their attention to these matters, according to Woodworth, eager "to be able to do for law what has been done for medicine by physiology. . . . They hope to substitute exact tests for vague general impressions."[59] The most important application of psychology was evidently the detection of false testimony. The professor described the word association/chronoscope method for determining guilt. The reporter then asked if "an infallible test could be evolved for detecting liars by means of their emotional expression?" Woodworth responded that although infallibility was unlikely because some people "actually believe the lies they tell" (as had been the case with Harry Orchard, as Münsterberg had discovered), eventually tests would indeed be able to catch the majority of deceptive witnesses. "All that psychologists hope to do is to make knowledge accurate where it is now inaccurate," said Woodworth. As the piece demonstrated, the most fascinating and newsworthy aspects of applied psychology concerned machines that could "detect a man in the act of telling a lie."[60]

These instruments articulated the dreams of applied psychology. A 1908 *Harper's Weekly* piece vividly articulated the fantasy: "One can imagine a witness giving evidence in court, while the soul machine records his emotions on a screen before the jury; and the conclusion will be drawn that the witness would be inclined to tell the truth rather than explain to the jury the reasons for the excursions of the galvanic wave."[61] There had recently been perfected one particular instrument, the author claimed, "which promises to achieve results revolutionary to our whole social system, our ethics and jurisprudence."[62] "It is a gauge of truth: in contact with it one cannot speak, even think, falsely without detection." Frank Marshall White's piece provided a more elaborate description of the "electric psychometer" than the *New York Times* article had some eighteen months earlier.[63] In addition to a detailed explanation of the galvanometer and the accompanying word association technique, the *Harper's Weekly* feature also printed photographs of "the inventor," Dr. Frederick Peterson, and "the machine at work." Recent experiments had proved that the electric psychometer—"already in practical use by neurologists—was possessed of limitless possibilities in the detection of crime and of false evidence on the witness stand. By its means the hidden thoughts of the subject may be reached, the secrets most carefully guarded in his innermost consciousness wrested from him, and his emotions measured mathematically. It is already in use to detect the lie the patient tells his physician, and it will doubtless be employed to detect the lie the criminal tells the police, the lie uttered by the perjurer in court." Detectives would have to abolish the coarse, brutal, and generally inconclusive methods of the third degree, and instead seek "scientifically accurate results by means of the more refined torture of the psychometer."[64] The article contained the first description of the questioning of a suspect in a manner that would become typical of the detection of deception, and it also used a sign that lie detection discourse would later find indispensable, the accuracy statistic: "in ninety-nine cases out of one hundred it would create an emotional complex that would register itself on the psychometer."[65] It even described a methodological innovation that polygraph operators would later use extensively: the sham experiment.[66]

The press encouraged scientists to claim that the new machines could discern mental facts by measuring bodily effects in order to diagnose morbid conditions. The newspapers were fascinated by McLane Hamilton's "mysterious little machine," Peterson's "soul machine," and Münsterberg's "machine for cure of liars." Spectacular knowledge claims found receptive audiences. Having emerged from the new psychology, the popularity of these "truth-compelling machines" was also partly a function of their ability to resonate with broader social concerns such as the fight against crime. Science and technology was to assist the quest for law and order. A brief notice in the *Scientific American Supplement* for 1909 was typical of the progressive sensibility: "Photography in the Service of the Law. The Scientific Detection of Crime."[67] The piece was illustrated with a picture of the "microphotographic camera" and a number of images taken with it. "These few examples," the article concluded, "may serve to give an idea of the important service which photography, in the hands of experts, is able to render to the cause of justice."[68] A reviewer for *Current Literature* predicted in 1911 that "the great scientist will supersede the great detective." Employing science to fight crime was doubly necessary now that criminals were arming themselves with the new scientific techniques. "The swindler and the murderer are proving themselves psychologists of power, chemists of great knowledge, electricians of genius. The great detective must meet the great criminal upon a plane of intellectual equality. He fails to do that nowadays, and this circumstance accounts for the relatively large amount of undetected and mysterious crime."[69] The training of the detective of the future would take place in the scientific laboratory. The caption to a photograph of a physiological laboratory at the Sorbonne suggested that a certain Professor Lapicque was "A Sherlock Holmes of Science." It was here that "the Paris police have more than once made tests that brought some evil-doer to destruction."[70] "In physical science the fundamental thing is laboratory experiment," agreed *The Literary Digest*, "and something of the same kind is necessary in the study of crime if we are to have trustworthy knowledge and permanent results."[71] Describing the workings of the galvanometer in *Harper's Weekly*, one physician admitted that although the mechanism by which the conscious will acted on the body was unknown, passion nevertheless had an emphatic effect on "heart, lungs, vessels, sweat glands, muscles, on all motile portions of the body. These things being so, why should not some instrument measure thought *by its effects on the body?* To be sure!"[72] Linking science and mystery, body and mind, and technology and humanitarian progress, the physician asserted that the galvanometer would "become exalted as a mind-reader, and 'sweat-box' interviews [would] take rank as scientific performances instead of will-breakers."[73]

Studying a criminal in his cell "mentally, morally, and physically, and with instruments of precision" constitutes a laboratory, asserted Arthur MacDonald, who wished to "turn our prisons into laboratories for studying the symptoms of evil-doers."[74] MacDonald's proposal is instructive because of the contrast between its own encyclopedic ambitions and the lie detector's humble efficiency. His ultimate target—the criminal degenerate—would be of no concern to the lie detector's advocates. MacDonald proposed that his painstaking study "would consist in a physical, mental, moral, and social study of each boy, including such data as age, date of birth, height, weight, sitting height, color of hair, eyes, skin, first born, second born, or later born, strength of hand grasp, left-handed, length, width, and circumference of head, distance between zygomatic arches, corners of eyes, length and width of ears, hands, and mouth, thickness of lips, measurements of sensitivity to heat and pain, examination of lungs, eyes, pulse, and respiration, nationality, occupation, education, and social condition of parents, whether one or both are dead or drunkards, stepchildren or not, hereditary taint, *stigmata* of degeneration."[75] Having written the first American treatise on criminal anthropology ten years earlier, MacDonald remained committed to Lombroso's personological theory of the born criminal. Because criminality was thought to be manifested in multiple ways, enormous efforts had to be devoted to discovering "all the corporal manifestations" of the criminal's being. But as

MacDonald described it, even though criminal anthropology's methods were laborious they could not guarantee results.

In 1911, a *New York Times* article was intriguingly titled "Electric Machine to Tell Guilt of Criminals."[76] "If It Is Perfected So As to Be Infallible," claimed the subheading, "It Will Make Expert Testimony Unnecessary and May Eliminate Juries in Trials." The iconography of the piece was revealing. Portraits of Professor Edward R. Johnstone and Dr. Henry H. Goddard were placed at the bottom of the page above a photograph of the psychometer. Two further views of the psychometer were placed below the headline. Connecting the five pictures was an illustration that would later play an enormous rhetorical role within lie detector discourse: the graphical record of the examination. "A Record of the Psychometer" was the caption; "Note the Disturbance in the Patient When the Word Whiskey Was Spoken." Sure enough, the tortuous vertical line indeed showed a disturbance—a high peak—following the utterance of the important word. The piece began with a futuristic fictional flourish: "May it please your Honor, the State accuses this man of the murder of his wife. I offer for your Honor's inspection these documents which prove beyond a doubt that the prisoner is guilty." "The documents are the records made by various instruments to which the prisoner was subjected in the State's psychical laboratory, and your Honor will find recorded there the incontrovertible evidence that this man committed the crime, the exact details showing every step taken prior to, during, and after the murder, the motives, and the attempts to throw suspicions upon others." The fantasy continued by directing the judge to "consult the record of the psychometer" for evidence of the prisoner's weakening determination: "a full confession will be forthcoming in a short time. In view of this conclusive proof the State asks for the maximum penalty of the law."[77] There was every reason to believe, the piece concluded, "that it will be a common thing for our grandchildren or our great-grandchildren to listen to just such arguments in criminal cases."

The writer was convinced of the psychometer's future role in the courts: "There will be no jury, no horde of detectives and witnesses, no charges and countercharges, and no attorney for the defense. These impediments of our courts will be unnecessary. The State will merely submit all suspects in a case to the tests of scientific instruments, and as these instruments cannot be made to make mistakes nor tell lies, their evidence will be conclusive of guilt or innocence, and the court will deliver sentence accordingly."[78] It is not surprising that great things were expected of it, given that "even in its present crude state no living man can conceal his emotions from the uncanny instrument." Nor was it implausible to imagine that the men who were "delving deeply into the most intangible of all things—human thoughts and emotions"—would "go down into scientific history." Although the piece had opened with a description of the instrument's potential legal impact, it was, however, predominantly a description of how the psychometer could be used to diagnose the "insane or weak-minded" thus benefiting "the mental health of the entire race." The article described the activities and ambitions of Johnstone, a leader of the American eugenics movement and Superintendent of the New Jersey Training School for Feeble-Minded Boys and Girls in Vineland, and Goddard, Director of Research. It was Goddard who explained to the reporter that the psychometer could be used to see whether or not a child with Down Syndrome "had any emotion in him or not." The presence of a female assistant or the sight of a piece of candy would apparently effect a reaction. The plan was to "take up the emotions one by one until we know just what departments of the feeble minds are dead and what are still alive. In that way we will have some basis for future psychologists to work upon in the hope of improving minds which to-day are hopeless even under the best treatment."[79] Despite describing the scientists as being "big-hearted" and "self-sacrificing," the article concluded on a sinister note: "if, a hundred years from now, there is an insane or a weak-minded person in all the world, it will not be the fault of Goddard or Johnstone."

The research at Vineland was motivated by a commitment to eugenics, the science of the managed selection of people according to their putative "fitness." Appointed director of research in 1906, Henry Goddard was "a superenthusiast of the eugenics movement," obsessed with halting the spread of feeble-mindedness.[80] Goddard demonstrated the connections between feeblemindedness and criminality in two books, *The Kallikak Family* (1912) and *The Criminal Imbecile* (1915).[81] These works of eugenics propaganda enjoyed a wide readership thanks to their richly mythopoeic storytelling qualities.[82] In 1908 Goddard had returned from a trip to Europe armed with a new method for measuring intelligence. The Binet-Simon tests he had encountered in Belgium suggested that the current system, which was based on medical classifications of feeble-mindedness—such as "microcephaly" and "Mongolism," could be replaced with a uniform scale of achievement based instead on performative abilities.[83] Intelligence tests promised to be able to identify the unfit with greater accuracy than had been previously possible, but they were criticized for being inaccurate and improperly administered. This is perhaps one reason why Goddard retained an interest in instruments such as the psychometer. Earlier in his career he had used an ergograph "to test will power," and an automatograph to measure involuntary motions.[84] One inventory of the training school's holdings lists the following pieces of apparatus: plethysmograph, pneumograph, ergograph, automatograph, dynamometer, and chronoscope.[85]

Goddard proposed that feebleminded criminality could be detected through the deciphering of stigmata or other physical means. Criminality was an invisible problem, one that could only be revealed through psychological testing and family genealogies. Secrecy, appropriately, was a major theme of Goddard's popular books, which foregrounded the difficulties involved in penetrating appearances to reveal underlying realities.[86] For a while after 1910 Goddard was the most influential criminologist in the United States, the national expert on the causes of crime.[87] By collapsing upper-grade feeblemindedness and moral imbecility into the single category of the "moron," he delivered the "coup de grace to criminal anthropology."[88] Even so, because of its commitment to defective human kinds, the Vineland Training School could never have been the place where the lie detector was invented, even if it did help to consolidate some of the conditions for the instrument's later emergence. The lie detector could not materialize in a setting where criminality remained conflated with feeblemindedness, degeneration, hereditary defect, or any other form of inherent psychopathology.

A paradigm shift was imminent. "There is No Criminal Type," the *New York Times* announced in November 1913 in a full page article: "In other words the criminal is a normal person, not markedly different from the rest of humanity who have managed to keep out of prison. In other words, there are in ministers and Cambridge undergraduates and college professors the making of pickpockets and thieves, as well as murderers and forgers."[89] The article was a detailed account of Charles Goring's *The English Convict*, the book that, according to the historian of criminality Arthur Fink, "was more decisive perhaps than any other factor in undermining belief in a criminal anthropological type."[90] Other factors challenging the notion of the born criminal included William Healy's intensive study of a thousand delinquents (1915), the psychometric testing of over a million American army recruits in 1917, aetiological findings produced by the case study method of psychiatry and psychoanalysis, and, as we have already seen, the development of a sociological approach to crime.[91] Fink dated the end of criminal anthropology in the United States to 1915, the year in which the term "lie detector" was first coined. By then, he concluded, the study of crime "had come a long way

"Composite Portrait of 30 Criminals. From Actual Photographs." Detail from " 'There is No Criminal Type,' A decisive break with the past: Dr. Charles Goring, deputy medical officer, His Majesty's Prison, London declares that 'the criminal is a normal person, not markedly different from the rest of humanity.' " *New York Times*, November 2, 1913.

from the time when the madman was indistinguishable from the criminal, from the time when it was held that the shape of the skull or of the brain determined criminal or non-criminal behaviour, from the time when it was believed that there was a fixed criminal anthropological type . . . when it was asserted that every feeble-minded person was a criminal or a potential criminal."[92] It was also in 1915 that a "Psychology Squad" was established to advise the New York Police Department on "how to differentiate mental defectives from crooks."[93]

The writers of scientific detective pulp fiction, a popular American art form during the first half of the twentieth century, had long been at ease with the idea that there was no such thing as an inherently criminal type. This set them apart from the scientists about whose work they were reading in the newspapers.[94] They had also drawn inspiration from the fin-de-siècle Gothic novels that had critiqued criminology's central claims. A considerable influence on the form and style of the scientific detective stories, Sherlock Holmes embodied the breakdown of the barrier between the normal and the pathological that scientific criminology had tried so hard to establish. Following Holmes's lead, other pulp fiction detectives quickly acquired personal scientific laboratories.[95] Science and technology considerably influenced turn-of-the-century dime novels like *Nick Carter Weekly,* for example, a serial that made extensive use of X-ray machines, photographs, telephones, and microscopes. No wonder then that it was from within scientific detective pulp fiction that the lie detector finally emerged.

During the ten years preceding the outbreak of the First World War, a variety of new scientific detective characters emerged in mass circulation magazines. Their personas evolved from professors and doctors through psychologists and onto criminologists, in many cases adapting the latest piece of scientific equipment toward crime-fighting ends.[96] Dr. John Evelyn Thorndyke, a medico-legal expert, whose adventures spanned 1907 to 1942, was the creation of R. Austin Freeman, an ear, nose, and throat specialist. A disavowal of the compromised and neurotic fin-de-siècle Gothic medical man, Thorndyke was often described as having a very un-Lombrosoian "symmetrical" face. He had little to do with the corrupt discourse of *homo criminalis*, standing out "like a tower of stone," erect and energetic, radiating dignity and force of character. His creator saw no reason to make his detective ugly or eccentric, both of which signified problems of the psyche.[97] Freeman's adversaries and colleagues were derived from his archetype: professional, dependable, inventive. His sidekick was the omnicompetent Nathaniel Polton, a laboratory assistant, chemist, and photographic specialist who could build anything from an astronomical clock to a microscope. Algernon Blackwood's John Silence (1908) was a "Physician Extraordinary"; William Hope Hodgson's Carnaki (1910) "a Psychic Investigator"; Max Rittenberg's Dr. Xavier Wycherley (1911) "a professional psychologist and mental healer who detects on the side."[98]

But it is Luther Trant, America's "first scientific detective," who warrants a special mention in any history of the lie detector. Created in 1909 by Edwin Balmer and William MacHarg, two Chicago newspapermen, Trant's success was attributed to "the tried and accepted experiments of modern scientific psychology." Luther Trant, his name an anagram of "Learnt Truth," "humps his muscles against statistical deviations and is enthralled by response times and correlation factors." Trant used galvanometers, pneumographs, plethysmographs, and chronoscopes to measure physical response to emotional stress and so unmask miscreants.[99] Balmer and MacHarg's first Trant story, "The Man in the Room," which appeared in *Hampton's Magazine* in May 1909, had Trant arguing that psychological testing must be applied to civil criminal cases and not left as laboratory demonstrations. This story features an early example of psychophysical testing of a nonpathological subject to uncover guilty knowledge rather than demonstrate inherent criminality. At the beginning, Trant praises his mentor, Professor Reiland, for passing on his knowledge of "the cardiograph, by which the effect upon the heart of every act and passion can be read as a physician reads the pulse chart of his patient; the pneumograph, which traces the minutest meaning of the breathing; the galvanometer, that wonderful instrument which, though a man hold every feature and muscle passionless as death, will betray him through the sweat glands in the palms of his hands." He had also been taught "how a man not seen to stammer or hesitate, in perfect control of his speech and faculties, must surely show through his thought associations, which he cannot know he is betraying, the marks that any important act and every crime must make indelibly upon his mind."[100] In this first story, Trant uses a "pendulum chronoscope," a voice-actuated device that measures the delay between a word spoken and the response of the testee.[101]

In "The Fast Watch" (June, 1909), Trant uses some banana oil and a galvanometer to break an alibi. "The Man Higher Up" (September 1909) was reprinted in both *Amazing Stories* (December 1926) and *Scientific Detective Monthly* (February 1930) during the brief recapitulation of the scientific detective genre in the early 1930s. The cover of the latter shows a worried-looking suspect ("The Boss") attached to the instrument while Trant and assorted white-coated scientists look on. Bearing a vague resemblance to Hugo Münsterberg, a German psychologist called "Professor Kuno Schmalz" helps Trant to use a plethysmograph and a pneumograph to detect guilt. It is significant that the chief suspect in this story—which also features the first appearance of the image of three simultaneous physiological traces on a chart—is a white-collar company boss, a criminal with no taint of psychopathic deviancy.[102] In "The Eleventh Hour" (February 1910), Trant uses the electric psychometer (referred to as "the soul machine") to expose a sinister oriental execution plot. The galvanometer is connected to a lamp while a mirror spectacularly projects the suspect's guilt onto the wall for all to see. Balmer and MacHarg were anxious to emphasize that the instruments they were describing, although "little known to the general public," were "precisely such as are being used daily in the psychological laboratories of the great universities — both in America and Europe — by means of which modern men of science are at last disclosing and defining the workings of that oldest of world-mysteries—the human mind." Experiments in university laboratories had evidently established that "the resistance of the human body to a weak electric current varies when the subject is frightened or undergoes emotion; and the consequent variation in the strength of the current, depending directly upon the amount of emotional disturbance, can be registered by the galvanometer for all to see." Luther Trant did not invent anything new, the authors explained, he merely adapted the accepted experiments of modern scientific psychology. He may have been "a character of fiction; but his methods are matters of fact." Implying that they had pioneered a new use for the technology, Balmer and MacHarg claimed that the instruments had previously been employed merely to diagnose madness: "If these facts are not used as yet except in the academic experiments of the psychological laboratories and the very real and useful purpose to which they have been put in the diagnosis of insanities, it is not because they are incapable of wider use."[103] As Trant put it, "I have merely made some practical applications of simple psychological experiments, which should have been put into police procedure years ago."[104] He "read from the marks made upon minds by a crime," he explained, "not from scrawls and thumbprints upon paper . . . But by the authority of the new science—the new knowledge of humanity—which he was laboring to establish."[105] "Instead of analyzing evidence by the haphazard methods of the courts," Trant explained, "we can analyze it scientifically, exactly, incontrovertibly—we can select infallibly the truth from the false."[106]

Serialized mass circulation magazines were tailor-made for the "whodunit" plot structure in which a number of suspects were initially considered equally likely to have committed a crime. Pretty young women, delicate and flawless in appearance, often became prime suspects, much to the surprise of onlookers. Criminal anthropology's crude concept of the born criminal was much less suited to this narrative format. In Arthur Reeve's first Craig Kennedy story, "The Case of Helen Bond" (December 1910), the suspect is described as "the ideal type of 'new' woman—tall and athletic, yet without any affectation of mannishness . . . her dark hair and large brown eyes and the tan of many suns on her face and arms betokened anything but the neurasthenic. One felt instinctively that she was, with all her athletic grace, primarily a womanly woman."[107] In "The Crimeometer" (December 1912), the psychologist-turned-detective was anxious to rule out hereditary defects such as epilepsy or "abnormal conformation of the head" before submitting the suspect to a truth test. The galvanometer "psychometer" is described as "an actual working fact. No living man can conceal his emotions from the uncanny instrument. He may bring the most gigantic of will-powers into play to conceal his inner feelings and the psychometer will record the very work which he makes this will-power do." In the Luther Trant story "Eleventh Hour" (February 1910), Balmer and MacHarg explicitly repudiated racial typologies with a plot that had Trant test four Chinese suspects. "You can't get anything out of a Chinaman! Inspector Walker will tell you that!" says an attending detective. "I know Siler, that it is absolutely hopeless to expect a confession from a Chinaman," Trant replies, "they are so accustomed to control the obvious signs of fear, guilt, the slightest trace or hint of emotion, even under the most rigid examination, that it had come to be regarded as a characteristic of the race. But the new psychology does not deal with those obvious signs; it deals with the involuntary reactions in the blood and glands which are common to all men alike—even to Chinamen!"[108]

Arthur Reeve created "Craig Kennedy, Scientific Detective," while he was still an undergraduate at Princeton University. He had read a series of articles of scientific crime detection while working as a journalist on *The Survey* (1907).[109] His aim for the character, he later recalled, was to combine "science and law in a Nick Carter who should have both the University and Third Avenue melodrama in his make-up."[110] The first Craig Kennedy series ran in *Cosmopolitan*, the magazine Münsterberg had also written for, from December 1910 to October 1912. It became enormously successful and ran in pulp fiction titles such as *Amazing Detective Tales* and *Scientific Detective Monthly* until 1935.[111] Reeves' plots featured a large number of suspects but very few in-depth character descriptions. His narratives often pivoted not on assaying the depth of the depravity of one particular criminal but rather on the question which, among many apparent innocents, was culpable.[112] The new technologies could assist the psychologically-informed detective to uncover the truth. In "The Truth Detector" (1910), Kennedy explained that the pneumograph "shows the actual intensity of the emotions by recording their effects on the heart and lungs together. The truth can literally be tapped, even where no confession can be extracted. A moment's glance at this line, traced here by each of you, can tell the expert more than words."[113]

In the first Kennedy story, "The Case of Helen Bond," the suspect is subjected to a deception test using a plethysmograph and a word association test: "She smiled languidly, as he adjusted a long, tightly fitting rubber glove on her shapely forearm and then encased it in a larger, absolutely inflexible covering of leather. Between the rubber glove and the leather covering was a liquid communicating by a glass tube with a sort of dial. Craig had often explained to me how the pressure of the blood was registered most minutely on the dial, showing the varied emotions as keenly as if you had taken a peep into the very mind of the subject. I think the experimental psychologists called the thing a 'plethysmograph.'" Reeve also equipped Kennedy with "a very delicate stop-watch" for measuring association time. "Neither of us was unfamiliar with the process," asserted Kennedy's sidekick, a newspaper

"The truth can literally be tapped," explained "Craig Kennedy, Scientific Detective," "even where no confession can be extracted." "The Truth Detector," 1910. Frontispiece, Arthur B. Reeve, *The War Terror* (New York: Harper & Bros., 1915). Digitized by UCSD.

reporter, "for when we were in college these instruments were just coming into use in America. Kennedy had never let his particular branch of science narrow him, but had made a practice of keeping abreast of all the important discoveries and methods in other fields. Besides, I had read articles about the chronoscope, the plethysmograph, the sphygmograph, and others of the new psychological instruments. Craig carried it off, however, as if he did that sort of thing as an every-day employment."[114]

The writers of detective pulp fiction accorded the technology a reliability the scientists were reluctant to condone. Fiction writers deployed an enormous range of technologies to detect deception, from reaction time to respiration rates and from galvanometers to pulsometers. Jacques Futrelle's detective used the simple technique of measuring pulse to unmask the culprit in his short story "The Motor Boat" (1906).[115] In Charles Walk's *The Yellow Circle* (1909), a character proposes giving a word association test to a butler he suspects of theft: "The chap must say the first word that pops into his mind, suggested by the word you gave him; the machine measures the interval of thought, and if there is nothing to interfere with the association of ideas, the chap will answer prompt the first word that your word suggests. Hesitation signifies equivocation."[116] In Arthur Reeve's story "The Lie Detector" (1915), Craig Kennedy described a "psychophysical test for falsehood" based on the inspiration-expiration ratio. "Lying," he explained, "when it is practiced by an expert, is not easily detected by the most careful scrutiny of the liar's appearance and manner." But successful means had been developed for the detection of falsehood by the study of experimental psychology. These effects were "unerring, unequivocal. The utterance of a false statement increases respiration; of a true statement decreases. . . . This is a certain and objective criterion . . . between truth and falsehood. Even when a clever liar endeavors to escape detection by breathing irregularly, it is likely to fail. . . . You see, the quotient obtained by dividing the time of inspiration by the time of expiration gives me the result."[117] Cleveland Moffett's (1909) novel *Through the Wall* featured one Dr. Duprat using a "pneumatic arrangement" to recognize the mental states of criminals, "especially any emotional disturbances connected with fear, anger or remorse."[118] "Every nation in the world has, at some time or other, had its ordeals for the detection of guilt," a lawyer tells a jury in Melvin L. Severy's novel *The Mystery of June 13th* (1905). "Several instruments may be used for this purpose, among them the cardiograph, kymograph, hemadynamometer, sphygmograph, sphygmophone and hematachometer."[119]

From around 1904 to the First World War, such instruments fascinated novelists, journalists, and scientists alike. A type of person invariably resided at the center of the scientific discourse: the insane patient, the habitual criminal, the eugenically unfit degenerate, the feeble-minded child. In Münsterberg's case the target was indeed a liar—but one in search of a cure for the deeper ills that rendered him untrustworthy. For the scientists the aim was to uncover the correlates of criminality within the criminal self, to assess the depth of depravity with a view to treating it.[120] The novelists, journalists, and writers of detective pulp fiction, however, were interested in the pragmatic and plot-driven question as to which, among many apparently innocent and ostensibly normal suspects, was guilty. Unlike criminal anthropology, which had amassed its resources around the most degenerate suspect, pulp fiction privileged the person that everyone least suspected. To a considerable extent, the lie detector was an invention of those fiction writers for whom the key issue was simply the presence or absence of guilt within one individual among many. The lie detector, that great twentieth century legend, was quintessentially a byproduct of the whodunit. It materialized not from the laboratory but from the story.

CHAPTER 6

"Some of the darndest lies you ever heard"

Who Invented the Lie Detector?

"There's a contrivance recently invented by some college professor," said he, "that I'd like to try on Cullimore. It is a lie detector; with its aid one can plumb the bottomless pits of a chap's subconscious mind, and fathom all the mysteries of his subliminal ego. You set some wheels going, the chap lays his hands on a what-you-call-'em, and then you proceed to fire some words at him. It is like a game."
—Charles Walk, *The Yellow Circle* (1909)

"What electric investigative device was invented by Nova Scotia-born John Augustus Larson in 1921?"
—*Trivial Pursuit* question (ca. 1996)

According to the popular general knowledge game *Trivial Pursuit,* John Augustus Larson invented the lie detector in 1921. The question appears to be simple, the answer clear-cut. The American press certainly considered the issue unproblematic: "The 'lie-detector' machine that records tell-tale changes in heart action and breathing accompanying deception," *Survey* magazine reported in 1929, "was invented by Dr. John A. Larson in 1921."[1] A 1922 *San Francisco Examiner* article was titled "Inventor of Lie Detector Traps Bride": "Dr. John Augustus Larsen [sic] . . . has lately emerged upon the stage of fame as the inventor of the sphyg—sphygomanom—call it the 'lie-detector.' "[2] Larson's lie detector was an "interesting device, with great possibilities" according to *The Literary Digest* in 1931, "yet even its inventor regards it as not yet perfected."[3] Reviewing the history and development of the lie detector in

1938, Larson himself tacitly confirmed that indeed it was he who had invented the instrument in 1921.[4]

Although the historical record thus provides some evidence to support *Trivial Pursuit's* contention, the game's question-setters nevertheless chose to privilege one candidate for inventor status over others. In a 1932 feature, "Science Trails The Criminal," *Scientific American* printed a photograph of "The designer, Mr. Leonarde Keeler, of the polygraph or so-called "lie-detector" giving a demonstration of a deception test."[5] In 1933, the *New York Times* reported that "Leonarde Keeler, 29-year-old inventor of the lie detector, "had been presented with an award for making a most outstanding civic contribution to Chicago.[6] The *Review of Reviews* praised Keeler for being "one of the first scientists to see the possibilities of the polygraph lie-detector," claiming that "no more important invention has ever been made for successfully dealing with crime in the whole course of criminal science."[7] Implying that the instrument was Berkeley police chief August Vollmer's innovation, however, *Outlook and Independent* in 1929 spoke of "the Vollmer pneumo-cardio-sphygmometer, or 'lie-detector.'"[8] In March 1937, the *New York Times* asserted that Leonarde Keeler was "the "inventor of the detector, scientifically known as a polygraph,"[9] but seven months later it named William Moulton Marston as the "inventor of the lie-detector."[10]

The historical record reveals numerous claimants to the title. As Marston perceptively remarked, in fact, there were almost as many inventors of the lie detector "as there were monks, in the old days, who claimed to possess a piece of the true cross."[11] During the early period of the machine's development, there was no consensus as to who had actually invented the device—although almost everyone agreed that the issue of invention was pertinent. The ambition of this chapter, however, is not to arbitrate between the various competing claims in order to locate the "true" inventor. Instead, it will address two fundamental issues. First, I argue that it is not legitimate to credit the "invention" of the lie detector to a single individual. Second, I explore how, despite being a myth, invention has nevertheless played a constructive role throughout the instrument's history.

If a lie detector is defined as an instrument used to record the physiological reactions of a nonpathological subject, then such an instrument was first described in Balmer and MacHarg's inaugural Luther Trant story, "The Man in the Room" (*Hampton's Magazine*, May 1909). The plot of this story is recapitulated in Arthur Reeve's first Craig Kennedy story, "The Case of Helen

Bond" (*Cosmopolitan*, December 1910). In both stories a young female suspect is subjected to a word association/reaction time test using either a "pendulum chronoscope" or a "plethysmograph." From these sources—described by one commentator as "blood relations, if not twins"—one set in Chicago, the other in New York, fact has followed fiction in the form of two parallel narratives that tell the story of the invention of the lie detector.[12] While the first begins in Berkeley, later moving to Chicago, with the work of John Larson, Leonarde Keeler, and August Vollmer, the second has its origins in Boston and features one man, William Moulton Marston. While the former is a tale of the practical concerns of a professionalizing police force, the latter is the story of an academic psychologist who gradually made a transition into the public domain.

In April 1924, *Current Opinion* reported on a "painless method of enforced confession."[13] Professor John A. Larson "of the University of California and consulting crime expert of the police department of Berkeley, California" had "perfected an instrument for 'nailing the lie.'" The idea was "based on the fact that under the excitement of questioning, heartbeats and breathing cannot be controlled." Another article about the lie detector appeared in *Collier's* magazine that August.[14] "The Future Looks Dark for Liars," it announced, "For Those Scientific Men Now Have a Lie Detector That Actually Works." The author, Frederick Collins, asked Dr. Charles Sloan of the *Los Angeles Times* to explain the workings of the instrument. It "is based on two well-known methods of registering human impulses," he said: the sphygmomanometer and the pneumograph: "That's the way the lie detector works, the way it is working in Chief Vollmer's campaign against Los Angeles's highly advertised crime wave." The machine's sponsors were Dr. Herman M. Adler, Dr. John Larson, and their "star pupil," Leonarde Keeler.[15] In spite of his achievements, Keeler was only twenty years old and still a student. "'The first model of the lie detector—I call it the emotiograph—was very crude,'" "the youthful inventor" explained. "It occupied a whole table six feet in length." *Collier's* attributed the creation of the instrument to the team of Larson, Vollmer, and Keeler. Berkeley Police Department chief of police August Vollmer was the most senior of the three. He was a tireless campaigner, an important influence on the transformation of the American police from a low-status, disorganized, and incompetent body of men, into an institution for which values of professional crime fighting and serving the community were paramount.[16]

In 1938 the psychologist William Moulton Marston offered an alternative

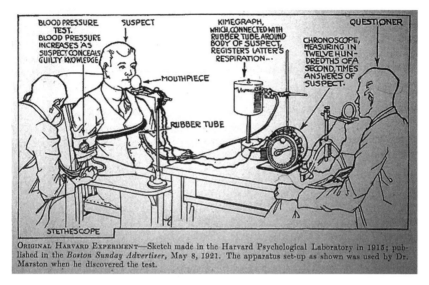

ORIGINAL HARVARD EXPERIMENT—Sketch made in the Harvard Psychological Laboratory in 1915; published in the *Boston Sunday Advertiser*, May 8, 1921. The apparatus set-up as shown was used by Dr. Marston when he discovered the test.

"New Machine Detects Liars," announced the *Boston Sunday Advertiser* in May 1921. "Registers Emotions Scientifically; Traps Shrewdest Criminals." Reprinted by permission of *Boston Sunday Advertiser* / Hearst Corporation.

account of the origins in *The Lie Detector Test*.[17] Marston wanted posterity to credit him with the development of the deception test, as his immodest entry in the *Encyclopedia of American Biography* evidenced: "the remarkable thing is that he discovered his 'Lie Detector' while still an undergraduate, while all the big psychologists of the world had been trying to get a practical test for deception for the last fifty years."[18] Marston cited a *Boston Sunday Advertiser* article from May 1921 to support his claim to priority. "New Machine Detects Liars," it trumpeted, "Registers Emotions Scientifically; Traps Shrewdest Criminals."[19] "Successful lying will soon be a lost art," the article began: "William Moulton Marston, Boston lawyer-scientist, inventor of the psychological lie-detector, which he put forward in 1913 and has since greatly improved, has already sprinkled the way of the transgressor with thorns from Massachusetts to California. No matter how accomplished at ordinary deception a man may be, he cannot hope to deceive Mr. Marston's apparatus any more than a woman can humbug a weighing machine by lacing tightly and dressing in black."[20]

In an autobiographical sketch for the *Harvard Class of 1915 25th Anniversary Report*, Marston claimed he had "had the luck to discover the so-called Marston Deception Test, better known as The Lie Detector" in 1915.[21] His author profile for a 1940 popular psychology article reported, "Dr. Marston has won fame as inventor of the so-called lie detector and a writer on psychological subjects."[22] His entry in *The National Cyclopaedia of American Biography* was also notably animated: "In 1915, he became an assistant in psychology at Radcliffe College. In the same year in the Harvard psychology laboratory he originated the systolic blood pressure test for deception, better known as the Marston 'lie detector' test. This test, which is said to be the only officially recognized lie detector test in the world, is based primarily on the theory that deception or untruth elevates the blood pressure of the person practicing it."[23] In addition to crediting himself with pioneer status, Marston diminished the importance of the work of others. While his entry in the *Dictionary of Psychology*, "Lie Detection," acknowledged Larson's work, it nevertheless asserted that the Berkeley physician "agreed essentially with Marston as to the reliability of the blood pressure test for deception if properly administered under controlled conditions by an experienced operator with suitable scientific training." "Larson made notable contributions to test technique and apparatus," it conceded.[24]

Narratives drawing on the Berkeley story invariably credit one or all three figures with the invention. This tradition ignores or minimizes Marston's contributions. Eugene Block's *Lie Detectors: Their History and Use* praised Marston only for achieving the first use of the instrument in a court of law.[25] Eloise Keeler's biography of her brother Leonarde similarly neglected Marston's work, asserting that the "original lie detector ... was the brainchild of Berkeley's famed chief of police, August Vollmer."[26] One historian of criminology reiterated the claim that John Larson had developed the first lie detector with Keeler's cooperation.[27] "Long before the 1920s, at a time when the lie detector was only a gleam in the eye of its inventor, Dr. John A. Larson, police officers were convinced a suspect would exhibit visible signs when he was lying," wrote the author of *Invisible Witness: The Use and Abuse of the New Technology of Crime Investigation*.[28] In his review of the history of "detecting the liar," Dwight G. McCarty discussed the work of Münsterberg, Larson, and Keeler, but failed to mention Marston at all.[29] Marston's name was also ignored by the author of *Police Professionalism*, who wrote that Vollmer encouraged "the development of an instrument for detecting the physiological changes that are associated with lying, known today as the polygraph."[30]

But Marston is credited with the invention in histories of comic books. The entry in *The World Encyclopedia of Comics* is typical: "Marston discovered the lie detector in 1915," it claims.[31] One historian of comics called Marston "the inventor of the lie detector."[32] Another reported, "one of [Marston's] most memorable accomplishments was the development of the lie detector."[33] A more recent celebration explored Marston's relationship with the lie detector but neglected to mention any of his contemporaries.[34] Another historian credited Marston with being "a key player in the development of the lie detector."[35] The casual observer could be forgiven for wondering if the lie detector was invented on the West coast by a team headed by an enthusiastic police reformer or on the East coast by a populist Harvard-trained psychologist.

Although most of the sources that attribute the invention of the lie detector to Vollmer and his "college cops" ignore or minimize the importance of Marston's work, it is clear that Vollmer himself was aware of the psychologist's early studies on the detection of deception. The "father of modern police science" was well acquainted with the literature in criminal law, criminology, and social science.[36] He was also the only police chief to be on the advisory board of the *Journal of Criminal Law and Criminology*.[37] By 1921, Marston had published three academic papers of potential interest to Vollmer: "Systolic Blood Pressure Symptoms of Deception" (1917), "Reaction Time Symptoms of Deception" (1920), and "Psychological Possibilities in the Deception Tests," the latter published in the *Journal of Criminal Law and Criminology* in 1921.[38]

Marston had concluded that a rise in blood pressure constituted a "practically infallible test of the *consciousness of an attitude of deception.*"[39] More specifically, he argued that "sudden sharp, short rises" of systolic blood pressure betrayed "substantial lies in an otherwise true story."[40] Blood pressure was measured intermittently, with a "Tycos" sphygmomanometer. The effectiveness of the test, Marston wrote, depended "almost entirely upon the construction and arrangement of the cross-examination and its proper correlation with the blood pressure readings, a system of signals between examiner and b.p. operator being necessary." It was also important, by inserting periods of rest and "questions upon irrelevant and indifferent subjects," to ascertain the discrepancy between the subject's normal blood pressure and "the fixed increase . . . due to the excitement caused by the test or by court procedure."[41]

Marston's apparatus consisted of not one, but three separate components, as the *Boston Sunday Advertiser* piece explained.[42] The chronoscope was used to test reaction times during the word association test. The resulting attribution of guilt was then confirmed by the kimegraph's breathing record: "The examiner examines the record of the machine. He finds that at every word pointed out by the chronoscope as a suspicious one, the suspect's breathing has shown a marked change. For, psychologists declare, a man breathes entirely differently when he is lying."[43] The sphygmomanometer showed whether "the suspect's blood pressure has mounted steadily during the cross-examination." "The more he lied the higher his blood pressure has climbed. The record shows what scientists call the 'lying curve.'" Of the three methods, the systolic blood pressure test was the most reliable, according to Marston. The effectiveness of the test, he asserted, depended almost entirely upon the construction and arrangement of the cross-examination and its correlation with the blood pressure readings. A system of signals between the examiner and operator was necessary due to the discontinuous measurement of blood pressure.

In his 1932 book, *Lying and Its Detection*, John Larson explained the difference between his and Marston's work was that whereas Marston used a discontinuous blood-pressure technique, he favored a continuous blood-pressure method.[44] He also maintained that deception might be indicated by a lowering of pressure, not necessarily a reduction, as Marston had claimed. Apart from the measurement of blood pressure to detect deception—which was not an innovation in itself—Larson accepted none of Marston's contentions. The Berkeley "college cop" saw no reason to measure the systolic blood pressure; he reasoned that a continuous reading would be more objective than a discontinuous one; and he surmised that a lowering of pressure might also signify deception.[45] Larson also employed measures of respiration and word association reaction time. Believing Marston's discontinuous method to be inadequate, Larson devised a test method for routine testing that, he claimed, "has remained unchanged whenever so-called polygraphs are used, the various changes being mechanical in character." Emphasizing that the key principles had been described in academic journals from 1921, Larson added that he also obtained a time curve with a chronoscope.[46] Because Larson's apparatus measured both blood pressure and respiration, he gave it the somewhat pedestrian title of "Cardio-Pneumo-Psychogram."[47]

While he was anxious to take credit for the innovation of continuous blood pressure measurement, Larson magnanimously attributed the creation of the lie detector to Marston. "The real 'lie detector,'" he wrote in the foreword to Marston's book, *The Lie Detector Test*, "is a test, a scientific procedure, originated by Dr. Marston in the Harvard Psychological Laboratory in 1915, and modified by me at Berkeley, California, beginning in 1921."[48] Recognizing the importance of "invention" in lie detector discourse, Larson equated the lie detector with Marston's procedures: "There is a mistaken impression abroad that a variety of 'lie-detectors' are in common use; in fact, that every operator who uses the one established test has 'invented' a 'new Lie Detector' of his own. This is not true. There is only one lie detection procedure thus far established, and I am glad to say that Marston and I agree essentially upon its fundamental points."[49] "The lie detector test must be used as Marston originally proposed it, and I developed its application in police investigation," Larson concluded, "as a truly scientific procedure administered and interpreted by scientifically qualified experts."

Although Marston later claimed that he had discovered the "Marston Systolic Blood Pressure Deception Test" in 1915, by 1921 he had yet to exclude the measurement of reaction times and breathing rates as significant indicators of deception. He had claimed that his systolic blood pressure deception test was the only scientifically recognized form of lie detector. But in 1921 he was still unsure which measure—blood pressure, respiration, or reaction time—exclusively signified guilt. In a research paper published in 1920, five years after his supposed discovery of the "only one lie detection procedure," Marston was using as many measures as practically possible for the diagnosis of guilt: "the practical value of psychological studies in this field lies almost wholly in a complete and comprehensive scientific discovery and analysis of all the psychological symptoms of deception rather than in attempted use of one isolated set of these symptoms for detection of deception on the part of witnesses or criminals."[50] Marston's position indicated that the lie still retained symbolic traces of the pathological liar as the exclusive target of interest. He was still using the interpretative resource of the human kind to explain why his methods sometimes failed: "Mr. Marston divides all liars into two groups—positive and negative. Positive liars are those who are normally truthful and who, when obliged to lie, respond with difficulty. Negative liars are talented prevaricators just described.... The reactions which he will give to the innocent and crucial questions will be exactly the reverse of the answers given by the normal liar."[51] The "negative type," Marston claimed, was "the gifted liar who would be expected to exhibit much less confusion were he lying than if he were telling the truth."[52] This type was betrayed by his rapid word association reaction times: "the only flaw in the negative type's efforts at deception lies in said efforts being *too successful*."[53] Marston was therefore still somewhat committed to the concept of inherent criminality. In 1921— six years after he claimed to have discovered the one scientifically reliable principle of lie detection—he was still using the word association/reaction time method together with an analysis of the symptomatology of psychological types. In fact, because he would later afford the lie detector therapeutic qualities, Marston would never completely repudiate his commitment to the earlier discourse of the "soul machine."

Larson and Marston's indebtedness to the word association method, a key element of the earlier discourse of the "soul machine," and their residual targeting of human kinds (the "feeble-minded" and other clinical cases for Larson; the "positive/negative type" of liar for Marston), indicates the importance of the historical context of any deception detecting technology. It is difficult to actually pinpoint an exact "founding moment" when the lie detector was supposedly invented because a lie detector is not merely a "machine"; it is a complex array of techniques, concepts, procedures, and symbols. By privileging the trope of invention, lie detector discourse foregrounds the hardware of instrumentation. The mysterious "black box" becomes a privileged focus of attention. The trope of invention also obscures the role that processes and techniques played in the emergence of the technology. By 1924, when magazines were beginning to credit Leonarde Keeler with the creation of the lie detector, he had stopped using the word association method, as the *Collier's* piece on the "young inventor" revealed.[54] The "invention of the lie detector" (if this phrase retains any credibility) was partly a function of the requirement that subjects had to simply utter "yes" or "no" in response to questioning, and were no longer obliged to make timed verbal word associations.

In a 1925 article, "Every Crime Is Entrenched Behind a Lie," *Scientific American* reported that according to the new scientific criminology, lying was the criminal's "first misstep."[55] "What I am proving almost every day," reported Edward Oscar Heinrich, consulting criminologist of Berkeley, California, "is that crime is entrenched behind a lie; puncture the lie and the criminal is disclosed." There followed a description of methods for exposing devious criminal practices such as photomicrography, ballistics, and chemical analysis. By the 1920s, psychometrics, statistics, sociology, and genetics had demolished the barrier criminal anthropology had erected between the normal and the abnormal. Criminology was now interested in deception on its own terms, rather than as a symptom of an underlying mental pathology. Up until the 1890s, under the influence of a biological approach to crime, criminology sought to "diagnose crime" or "detect the criminal."[56] The facially disfigured criminal of the 1890s had, by the 1920s, escaped from the clinic and returned to the population to become but a faceless and anonymous member of society.

A related development was the innovation of the term "lie detector." Apparently first used in Charles Walk's novel *The Yellow Circle* (1909), by 1922 the *San Francisco Examiner* noted that the term was a matter of common parlance: "Everybody has heard of the 'lie detector.' Put it on a criminal's arm and ask him a rude question and if he lies the little wings of the machine will flop up and down."[57] It took a few years before the term became fixed in the popular imagination, however, because similar terms remained in circulation. A brief notice in the *New York Times* in June 1922, for example, described a "lietector."[58] "Capillary electrometer" and "emotiograph" were also used.[59] Larson called his instrument the "Cardio-Pneumo-Psychogram," or the "emotion recorder."[60] Keeler initially gave his instrument the unwieldy title "Pneumo-Cardio-Sphygmogalvanograph." As the lie was beginning to replace the inherent criminal as the essential problem for criminology, it followed that the lie detector would come to embody the dream of criminology.

Even at this early stage in the machine's development, a number of confusing issues remained. Although the term "lie detector" would eventually triumph over competing terms, no single person would outsmart his competitors for the title "inventor of the lie detector." As the lie detector's fame grew and other instruments were developed, several other claimants to the title "inventor of the lie detector" emerged. In January 1937, for example, the *New York Times* reported that a "new type of 'lie detector'" was "the invention of the Rev. Walter G. Summers."[61] A 1938 *Newsweek* headline read "Lie Detection: Device Invented by Priest Wins First Court Recognition."[62] The "Rev. Walter G. Summers, S. J." maintained that "Science Can Get the Confession."[63] Father Summers had built an instrument based solely on the resurrected principle of psychogalvanometry. "Particularly enthusiastic about his invention," reported one newspaper, "he believed innocent parties could in the future be saved from prolonged questioning or even trial through elimination by the detector."[64] Whenever a newspaper credited use of the instrument with solving a crime or a popular magazine exhibited a novel type of lie detector, the article was invariably accompanied by a description of the instrument's inventor.[65] In January 1938, a *Look* magazine article described "A Machine To Measure Lies."[66] "The new thought wave lie detector is given a workout by Rosemary Price," said the caption to a photograph of a woman whose head and hands had been attached to two electrodes: "One contact is placed at the base of the brain and two more on the hands. Dr. Orlando F. Scott, the inventor of the machine, says that women respond with so much electrical energy that their lies are easier to detect than those of men."[67]

"There is a mistaken impression," complained Larson, "that every operator who uses the one established test has 'invented' a 'new Lie Detector' of his own."[68] "Magazine and newspaper articles often referred to Nard as the inventor of the lie detector," recalled Eloise Keeler: "Whenever this was called to his attention, he would explain he had developed his own instrument, the Keeler Polygraph (Greek for 'many pictures'), a neat, compact, portable instrument that he'd improved over the years. He also started the first school for polygraph examiners and was a pioneer in the field of lie detection."[69] Marston also scorned the notion of invention, like his mentor Münsterberg before him, preferring to be described as the "discoverer" of the systolic blood pressure deception test. "I have never claimed credit for the 'invention' of a single instrument in connection with the Lie Detector Test," he wrote, somewhat disingenuously.[70] He complained about those newspapers and magazines that erroneously described the instrument as an invention, and he listed those cases "likely to befuddle the popular mind about the fact that the Lie Detector is a test and not a machine, and also about the minor matter of who 'invented' it."[71] "Whoever happens to be using the Lie Detector Test at the moment," Marston concluded, "becomes its inventor, and whatever form of the old apparatus he happens to be using must be called a 'new Lie Detector' or a 'marvelous new machine' which has just been invented."[72] The psychologist astutely recognized that "old apparatus" had simply acquired a new use.

The "invention of the lie detector" was predominantly a matter not of technological advance, but rather of conceptual, procedural, and, to some extent, terminological innovation. Because its history consisted of relatively small changes made over a number of years by many different persons, it is therefore not legitimate to speak of the instrument as being an "invention" at all—at least in the way that term is popularly understood. The lie detector was merely exploiting a mythic tradition of invention that had been constructed during the age of industrialization following the Civil War. The period 1865 to the First World War was characterized in the United States by rapid urbanization and spectacular economic growth. Exports rose from $1.49 million in 1900 to $2.3 billion in 1914. America had become the world's leading producer of coal by 1913, and by 1919 it was supplying two-thirds of the world's oil.[73] The relatively high cost of labor encouraged investment in mechanization in the hope that industry could be made more efficient. Invention was the motor of progress and inventors the heroes of the age. This was the era of Alexander Graham Bell, Thomas Edison, Wilbur and Orville Wright, and Henry Ford. Their inventions—the telephone, the electric lamp, the motion picture, the airplane, and the motorcar—signified the very essence of modernity.

Popular ideology depicted these inventors as heroic figures.[74] They stood for the American ideal that through endeavor and perseverance the ordinary man could single-handedly produce great technological advances.[75] Behind the ideology of individualism lay the imperatives of capital: invention was fuelled by economic necessity. Edison patented over one thousand inventions while he was the nominal head of an enormous research factory. Henry Ford's mass production line had been made possible by a whole series of precision inventions.[76] The Wright brothers were heavily indebted to numerous earlier workers. Guglielmo Marconi amalgamated the work of others into a practical package. Such technological innovations as the incandescent light bulb, the airplane, and the automobile might well be privileged in the popular technological imaginary, but their existence is heavily indebted to intricate labor processes and social networks.[77] While nineteenth-century technology was dominated by invention, in the twentieth century, the history of science and technology was characterized more by innovation, the creation of novel commercial products.[78] This is certainly true of the lie detector, whose constituent parts—the sphygmomanometer, the pneumograph, the galvanometer, and the kymograph—were all developed in the nineteenth century. It was Leonarde Keeler who first brought all three components together in a single box and who took out a patent on a minor innovation dealing with the measurement of blood pressure.[79] The lie detector pioneers, like the generation of inventors that preceded them, both actively sought fame and exploited the heroic persona of the inventor that was already deeply rooted in American culture.[80]

It is ironic that at the very moment the lie detector was being promoted as a heroic invention, the notion of invention was itself being criticized by contemporary sociologists and historians.[81] Even today, it seems, it is widely believed that inventions are the result of revolutions brought about by heroic individual geniuses. This was the idea self-consciously promoted by the lie detector's advocates, a process assisted by concealing earlier technologies, obscuring their different ambitions, and playing up the instrument's mystique.[82] The instrument's history is considerably more complex than traditional heroic narratives allow, particularly those that focus on mundane details of hardware at the expense of the many social, cultural, conceptual, and disciplinary factors that rendered the project of lie detection plausible. Because such histories are committed to the idea that the lie detector was invented, they are obliged to choose which narrative tradition to privilege. If the lie detector was an invention, then there can be only one inventor, so the reasoning went.[83] This partly explains why the two narrative traditions, Berkeley and Boston, remain relatively independent.[84]

It also explains the curious and notable fact that the lie detector's principal actors mistrusted and disliked each other intensely. Larson and Keeler were not on convivial terms, according to Fred Inbau: "Larson thought that Keeler had stolen some of his ideas and whatnot, and that the instrument that later was developed and labeled the Keeler polygraph was really something that Larson was primarily responsible for. So there was that ill feeling."[85] There are a few references to Keeler in Larson's *Lying and Its Detection*, and certainly no "slanderous" ones, but Keeler nevertheless felt he had been betrayed.[86] "I received a copy of Larson's book the other day," he told his mentor August Vollmer in September 1932, "and rather felt he outdid himself in knocking me every time he had a chance. In some of the cases he cited, he obtained his information from foreign accounts; instead of coming to me for details and getting a true account of the facts, he published every slanderous thing he could think of about me. Of course, all this was put in after I read the original manuscript." Keeler regretted that Larson had to resort to publishing his criticisms of him, "rather than telling me about them to my face. I feel that I can hardly trust him in the future."[87]

Larson was "still up to his old tricks" two years later. Keeler had apparently done his best to be amicable and "to cooperate with him whenever possible": "He always seems so friendly in my presence, but behind my back that's a different story. To individuals and in public talks and articles, he slams and pokes and tells some of the darndest lies you ever heard. Now he has the delusion that I'm trying to get his job."[88] "If I thought we could save him from a complete mental breakdown," Keeler continued, "I'd give him a half interest in the patents, but I'm afraid even that wouldn't help." Keeler was even less impressed with Marston's *The Lie Detector Test*. Marston praised Keeler for being a worker "of high standing and accomplishment" although he did suggest that his apparatus might be "somewhat expensive."[89] "Great credit for developing systematic lie detection in the banks should go to Mr. Leonarde Keeler," Marston wrote, adding, "He has done pioneer service in applying the deception test to commercial uses."[90] On the whole, the "Boston lawyer-scientist" appears to have given Keeler his rightful dues, crediting him with having a "clever, persistent drive."[91] Marston was certainly keen to emphasize that Keeler had not invented the lie detector. But it was Marston's claim that the younger man had once been "over-zealous" in his efforts to secure one particular confession that had so enraged Keeler.[92] "I was, of course, extremely disappointed in Marston's book," he told Vollmer in 1938, "but can well understand that when Marston and Larson get their heads together my stock goes down 100%." "He not only took all the below-the-belt blows at me that he could but in many respects was extremely unfair in criticizing our work and myself because of newspaper accounts. He is apparently so small that he didn't have the decency to get proper statistics from me regarding the bank work we have done, and instead took a second-hand figure quoted by some Chicago bank official. But such is life; and when my book is ready for publication I am going to cite Marston and Larson only from their writings and am not going to stoop to using a lot of irrelevant adjectives in order to belittle them. I think that as a scientific account of the detection of deception the Marston book ranks zero." Keeler told Vollmer that his loyal acolytes Fred Inbau and Al Dunlap were both writing reviews of Marston's book, "and in both cases Marston will find them rather uncomplimentary."[93]

Inbau didn't disappoint, calling Marston's book "an Alice-in-Wonderland journey through the fields of lies, liars, and lie-detectors."[94] Inbau's complaint was that the monograph was not scientifically credible. In making such "sweeping statements" as "there is no reason why 90 per cent of society's crime bill cannot be written off by the Lie Detector," Inbau accused Marston of forgetting that "he is supposed to play the role of a *scientist* and not that of a popular magazine writer, or newspaper reporter, or special guest on some advertiser's radio program. Apparently the author has repeated such exaggerations so often in *Esquire,* in *This Week,* and over the *Nash* and other radio programs that he really believes them himself."[95] Having ridiculed Marston's claim to priority, Inbau suggested that the psychologist's attacks on other workers in the field came "close to being libelous." The book was "practically useless as a guide to a person who seeks to detect deception by means of a 'lie-detector.'" "It is not possessed of sufficient interest to warrant the attention of the average reader who seeks an insight into the catacombs of 'science,'" he continued. "And because of the manner in which it is written—with its self praise unconcealed beneath a guise of bashful modesty, with its unjustifiable attempts to discredit the efforts of other workers, and a denial of proper credit to the author's predecessors—it can only bring ridicule upon the subject matter and disrespect for its author."[96]

Inbau received a dose of his own medicine, however, when Larson took the opportunity to review his *Lie Detection and Criminal Interrogation*.[97] "Unfortunately space does not permit an exposition of the inaccuracies and inconsistencies of this book," Larson began. "The glaring mistakes and the lack of familiarity with basic physio-psychological and pathological principles involved can be understood when we remember that the author is primarily trained in law and has no degrees either in physiology, medicine or psychology." As Inbau had charged Marston of failing to recognize the contributions of others, so Larson accused Inbau: "he shows pages of charts with involved, inaccurate interpretations, and then presents garbled formulations in an attempt to explain a technique which is incorrect as described. He is apparently unfamiliar with the scientific literature, or he deliberately ignores key contributions which have scientific priority and arbitrarily assigns credit at will."[98]

Although they had once been "close as brothers," some rancor even developed between Keeler and Inbau.[99] Inbau later recalled that Keeler "became a frustrated individual and began to look with some degree of envy on the success I was achieving in case investigation and with my writing."[100] Keeler was particularly resentful when he returned from a Caribbean cruise to find Inbau "in the limelight" having solved a murder case in his absence.[101] Keeler's "ill feelings . . . were really greatly augmented" in 1938 when the crime laboratory was sold to the Chicago Police Department, and the younger man was taken on as director. "The police department, for a variety of reasons that were never fully revealed to me, did not want Keeler. . . . Keeler, having developed some paranoia about me, thought that I had planned for this all along. In any event, he thought I had engineered this whole thing, which had no factual basis whatsoever, and that I can document."[102]

Larson's disliking for Keeler extended beyond the latter's death. In 1951, by then the Superintendant of the Logansport State Hospital in Indiana, Larson wrote Vollmer to keep him "abreast of some developments in the field of deception testing."[103] The field was "chaotic," Larson complained; "everyone is talking about the so-called 'Lie-Detector.'" Larson's principal criticism was that Keeler and other "quacks" had commercialized the technology before it had been properly tested in clinical settings: "Keeler, Lee and many others with their polygraphs, psychographs, photopolygraphs, pathometers etc. have focussed their attention upon the sale of apparatus and the training of laymen whose background was inadequate." No scientific interest had been shown by most of the apparatus promoters, "including Lee and Keeler." Larson was particularly anxious to clear up any confusion regarding "the pioneer Berkeley work" that he had done. Keeler had neither "perfected any polygraph or improved any of the original technique as I used it." Writing about events that had happened thirty years earlier, Larson's report to the seventy-five-year-old "Chief" was curiously detailed: "If you recall, just before the first test I had been ill at home.... I then went to you, told you of Marston's article, and asked your permission to conduct the tests in the University of California laboratory, and outlined my proposed methods. You agreed, and we all know the results.... Keeler was at that time a high school boy in short pants." Larson was still annoyed that he had given Keeler credit as a collaborator on *Lying and Its Detection*, "even though he never even read the manuscript and contributed nothing to it." "So you see," Larson concluded, "that I, *in your Police department, encouraged by you,* was a good many years ahead of anyone else in using the "Lie Detector" in police work."[104] Even Larson, who disliked the term, wanted to go down on record as the man who had invented the lie detector.

It might be predicted, perhaps, that an enterprise that ostensibly sought truth would have encouraged suspicion among the principal actors. Nevertheless, the extent to which mistrust features in lie detector discourse is remarkable. Anxious to avoid engaging with each other in public, in private the lie detector's architects revealed their intense disrespect for each other's work. They criticized each other's questionable ethics, illegitimate qualifications, and grubby ambitions. Yet they also recognized that by engaging in a competition for the title "inventor of the lie detector" they undermined not only the claim that the lie detector was an invention at all but also that the credentials claimed on behalf of the instrument were scientifically valid. Although they each had different ambitions (and regardless of their own views as to whether the instrument deserved the invention epithet), Larson, Keeler, and Marston all believed that being known as the "inventor of the lie detector" was a prize worth fighting for. The title conferred status and fame on its holder, and held out the possibility of great wealth. But because in the popular imagination an invention could only boast a single inventor, competition for the title was fierce.

Even astute critics occasionally fail to see past the entrenched myth of the instrument's invention. As we have seen, Carl Jung wanted to use the galvanometer in order to comprehend the criminal's emotional "complexes of ideas," not to detect the lie. Nevertheless, one historian of psychodynamic psychology claimed that Jung's work "culminated with the invention of the lie-detector."[105] Hugo Münsterberg's biographer similarly alluded to his role in the creation of the polygraph.[106] Two sympathetic scholars claimed he had administered a battery of lie detection tests to Harry Orchard in 1907. The case brought him spectacular publicity "and the idea of a lie detector was

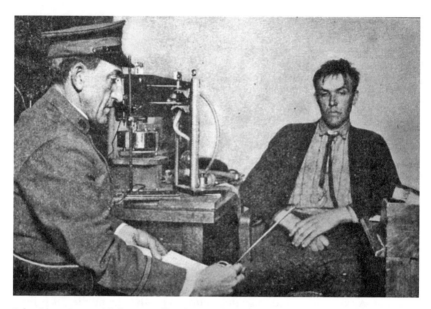

John Larson resisted the term "lie detector" but nevertheless wanted the credit for inventing it. John A. Larson, *Lying and Its Detection: A Study of Deception and Deception Tests* (New Jersey: Patterson Smith, 1932). Reprinted by permission of Patterson Smith Publishing, www.patterson-smith.com.

launched."[107] The main weapon in this "battery" of tests was the word association test. But Münsterberg's description of how he interpreted the results of the test indicated that he was still committed to a pathological theory of the criminal.

The question who invented the lie detector is problematic because the answer one gives depends on what one means by the term "lie detector." If the heart of the lie detector is taken to be the discontinuous technique of recording blood pressure then the credit is Marston's. If the essence is taken to be the continuous technique then the credit is either Larson's or Arnold Gesell's.[108] Yet as late as 1932 John Larson was insisting that the word association technique "must still remain a police tool—a very efficient police tool."[109] If the fundamental innovation is taken to be the assumed link between blood pressure and criminality then Mosso deserves recognition. In this last case, however, the relationship posited between blood pressure and criminality was quite different due to the different aims of the technology. And even if it were possible to decide on the basis of a single criterion of invention, there would still remain the problem of both Larson and Marston's residual commitments to clinical case studies as legitimate objects of knowledge. And what of Leonarde Keeler's claims to inventor status? After all, it was Keeler who designed a compact and portable lie detector.[110] It was Keeler who attempted to patent an instrument mechanism in 1925. And it was he who refined and stabilized the software of interrogation procedure.[111] In fact, none of the lie detector's so-called inventors can be credited with making any notable, imaginative, or technological innovations. Journalists, novelists, and writers of detective pulp fiction made the principal conceptual shift that made the lie detector possible—namely the repudiation of criminology's notion of the born criminal.

Nevertheless, from the early 1920s, the notion of "invention" permeated accounts of the instrument in newspapers, magazines, and books. The construct was necessary precisely because the lie detector was a new blend of old wines in a new bottle, old technologies applied to a new end. That the machine was depicted as an invention boasting an inventor gave it a reputation of scientific credibility. Such a controversial idea as "detecting lies" would hardly have been acceptable had the lie detector been described as a sphygmomanometer, a pneumograph, and a galvanometer housed in a box and operated by a technician. This is also why the "invention" of the lie detector was partly a function of the invention of the term "lie detector," a form of linguistic "black boxing": the simplification of scientific complexity and human agency.[112] "Invention" gave the lie detector credibility via a culturally valued origin myth, a numinous narrative that served to obscure the instrument's social and historical origins.[113] The myth of invention was primarily promoted in mass culture: national and local newspapers, the sibling-penned hagiography, popular psychology books and magazine articles, comic book histories, personal memoirs, pulp fiction. Although Marston objected to "picturesque press embroidery," and Larson complained that much of the material concerning lie detection was "being dished out through rewrite men to various magazines and newspaper reporters," both were reluctant to acknowledge the extent to which these domains were establishing the lie detector's very intelligibility. The untenable notion of invention—despite its lack of explanatory sophistication—nevertheless played a crucial role in the construction of this intelligibility.

CHAPTER 7

"A trick of burlesque employed . . . against dishonesty"

The Quest for Euphoric Security

"Myth . . . could not care less about contradictions so long as it establishes a euphoric security." —Roland Barthes (1972)

"Any kind of ordeal as a means of evidence is, of course, a derivative of charismatic justice." —Max Weber (1922)

The first accounts of the lie detector's history emphasized the scientific credentials of the instrument's advocates. Above all, they depicted the machine as being a product of modernity, the opposite of superstition. "In China there existed a practise of requesting an accused to chew rice and then spit it out for examination," wrote Fred Inbau in 1935 in *The Scientific Monthly*, "and if the rice were dry the subject was considered guilty."[1] "In India the movement of a suspect's big toe was supposed to be an indication of deception." "The interrogation of criminal suspects may not be easier today than formerly," wrote Paul Trovillo in his 1939 "History of Lie Detection," "but it is at least on a more objective basis."[2] Trovillo's history included accounts of the Ayur-Veda (900 BC), Erasistratus, the Greek physician (300–250 BC), ordeals of the Middle Ages, and the researches of Mosso and Lombroso. Lie detection "should be considered the fruit of centuries of germination," he wrote, "some of which, indeed, was plucked before it was ripe."[3] Historical accounts generally claimed that while the need to detect the truth had been a recurring problem, it was only in the early years of the twentieth century that science had finally solved it. William Moulton Marston, for example, opened his *The*

Lie Detector Test with a characteristic flourish: "Since that disastrous day in Eden when the subtle serpent said to Eve, 'Eat the fruit, for surely thou shalt not die,' lying has been the rule in human behavior and not the exception."[4] These accounts served an important function within lie detector discourse. First, they emphasized that the instrument was a modern scientific invention. Second, they deflected attention away from a fundamental truth of lie detecting: that the technology functioned according to one of the oldest methods of detecting deception available—the scrutinizing of body language, facial expression, words, and gestures. The origin myth of the lie detector thus contains a subtle falsehood.

From its emergence in the early 1920s, the lie detector was portrayed as the "antithesis of 'third-degree.'"[5] "It is a far cry from the rack and torture chamber of the Middle Ages to the modern 'Lie Detector,'" enthused *Current Opinion* in 1924.[6] A 1937 issue of the picture magazine *Look* contained a condemnation of the brutal practice together with an explanation of how the instrument worked.[7] "Prying open the stubborn jaws of the criminal for evidential data has always been one of the major tasks of the law," wrote Henry Morton Robinson, writing in *Forum and Century*, "and one that has rarely been accomplished without an accompanying taint of human dishonor and cruelty": "The rack, the pincers, and more recently the blackjack and rubber hose have been the favorite instruments used in this important legal process of 'snatching the confession.'" Six hundred years of Western civilization had evidently not halted the use of torture and coercion in wringing guilty knowledge from criminal suspects. Robinson contrasted the brutalism of entrenched police practices with science's humanitarian ambitions. He alerted his readers to a number of devices that had recently proved "tremendously successful in plucking desired truths from guilty lips." Although not yet in general use, they were nevertheless "clearly destined to supplant the barbarous torments of the third degree."[8] "Perhaps the most dramatic and satisfactory of these instruments," Robinson continued, "is the Keeler Polygraph, commonly known as the lie detector."

The lie detector emerged amid mounting criticism of the brutalities of the third degree. Progressive police theorists such as August Vollmer, together with periodicals such as *The Nation* and *The New Republic*, led the strident campaign to oppose the practice, a campaign that reached a crescendo in the early 1930s.[9] Emanuel Lavine's (1930) *The Third Degree: A Detailed and Appalling Expose of Police Brutality* was little more than a collection of sensational anecdotes. Published the year before the Wickersham Commission's

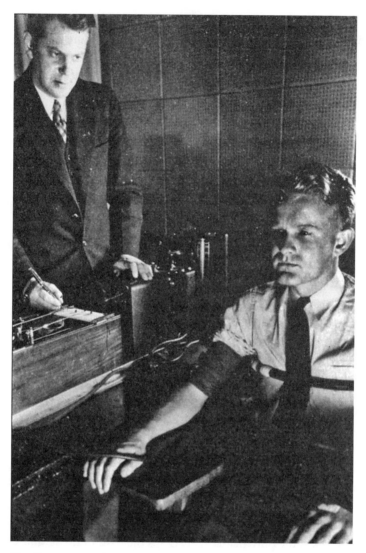

The dramatic air of the lie detector test was increased by the presence of a mysterious looking "black box." Eloise Keeler, *The Lie Detector Man: The Career and Cases of Leonarde Keeler* (Boston: Telshare Publishing, 1984). Reprinted by permission of Telshare Publishing Company.

official investigation into police lawlessness,[10] it nevertheless captured the public opposition to the practice. One author of the Report concluded that the regular and systematic use of violence to extract confessions was habitual in only the two largest cities of the United States. But if the third degree was defined as "threats, lies, display of weapons, exhausting grilling, and the like" then it was the exceptional American city where it was not practiced.[11]

Condemnations of police violence had been voiced from the earliest years of the century.[12] "Decent public opinion stands firmly against such barbarism," wrote Harvard psychologist Hugo Münsterberg, "and this opposition springs not only from sentimental horror and from aesthetic disgust: stronger, perhaps, than either of these is the instinctive conviction that the method is ineffective in bringing out the real truth."[13] Münsterberg's student William Moulton Marston later asserted that the use of deception testing exercised "a control over the criminal mind which no 'third degree' [could] produce and which those who know criminal psychology fully understand."[14] "The process is operated with the utmost consideration for the individual," remarked the *Review of Reviews* in 1932. "Third degree methods are discarded; the subject is given a cigar or cup of coffee if he desires."[15] The nation's progressive police were quick to recognize the value of Keeler's Polygraph, wrote *Scientific American*. "The need for such a device was acute. The old so-called 'third degree' methods of obtaining the truth had been proved inefficient."[16] *Saturday Evening Post* ridiculed the Chicago policeman who joked about having "'a lie detector of my own about this long' . . . indicating the length of the blackjack."[17] "As a matter of fact," the *Post* concluded, "the lie-detector examination is the opposite of the third degree."[18]

The interest of the popular press in the "lie-detector" exploded from around 1930. The instrument was so famous by 1936 that *Popular Science Monthly* could describe "Two Simple Ways to Make a Lie Detector" to its readers.[19] In some quarters, by the midthirties the instrument was synonymous with the "Keeler Polygraph." Articles on the lie detector in the popular press invariably attributed inventor status to an individual or simply asserted that the instrument was an invention. But if the lie detector was to be depicted as an impressive scientific invention, then its validity had to be expressed in the most scientific manner possible: by deploying the power of numbers. Articles about the instrument's accuracy and reliability invariably included a figure expressing either the rate of confession or the rate of lie detection. Although few methodological or technological innovations were made to the test during the 1930s, the accuracy figure gradually increased as the decade wore on. For a technology abounding in hubris, hyperbole was the norm.

In 1932 the *Review of Reviews* reported that confessions have been obtained in more than eighty-five percent of the cases on record.[20] In experimental cases, claimed the *Scientific Monthly*, "there is an accuracy of approximately 85 per cent." In numerous criminal cases however, "full confessions have been obtained in approximately 75 per cent of those in which the record indicated deception regarding the pertinent questions propounded of the suspect."[21] *Living Age* ignored the distinction between experimental and criminal cases when it claimed the ingenious machine had registered accurate results in "75 to 86 per cent of cases submitted to it."[22] *Forum and Century* claimed, "in seventy-five per cent of cases the subject, on viewing the unanswerable evidence he has given against himself, breaks down."[23] *Reader's Digest* enthused that "75 out of every 100 individuals who give guilty reactions on the lie-detector make full confessions after they are shown the polygraph's remorseless record of their prevarications."[24] *Newsweek* reported Keeler's belief "that on the average his machine is 85 per cent accurate in detecting lies to questions unimportant to the individual. Seventy-five per cent of all criminal cases found guilty by the apparatus have been substantiated by subsequent confessions."[25] After admitting some initial skepticism about the powers of the "Fordham psycho-galvanometer," New Jersey Assistant Prosecutor Edward Juska claimed he was "now convinced that the detector is 98 per cent correct, as claimed."[26]

It was no wonder the lie detector was considered to be a cost-effective addition to any police department. In Wichita, Kansas, the police apparently obtained approximately twenty confessions a month that would have not been secured without the use of the instrument. Statistics showed that approximately sixty percent of those caught lying by the polygraph confessed.[27] O. W. Wilson, a graduate of Vollmer's "college cop" program, headed the Wichita department. *Scientific American* called Wichita "the proving ground for the Keeler Polygraph," because apparently more people were tested there than anywhere else. Having tested more than thirteen hundred persons a year since 1936, the department had accumulated some four thousand recorded examinations by 1939—nearly half of which were the records of transients or vagrants, "picked up in the railroad yards or found loitering about the city streets." The writer produced a statistic that would eventually achieve notoriety within the detection of detection discourse: "Otherwise stated, 99.9 percent of all persons examined were able to produce records upon which a definite and immediate decision could be made."[28] In 1941, *Reader's Digest* claimed that the lie detector in regular use for six years in a large department store had "caught 90 per cent of the guilty, [and had] never convicted the innocent." An official of a large detective agency said that while half of its cases could be solved by direct investigation, detectives ran up "against a stone wall" in the other half. "Here they have found the Polygraph 99 percent perfect."[29] *Saturday Evening Post* concurred, maintaining that in only "1 per cent of the cases, the lie detector findings are proved to be wrong."[30]

Accuracy statistics were also privileged in the scientific literature. In his 1917 paper on the "Systolic Blood Pressure Symptoms of Deception," Marston claimed that measuring blood pressure constituted "a practically infallible test of the consciousness of an attitude of deception."[31] In one experimental study he concluded that he had correctly judged 94.2 percent of the curves he had examined.[32] Harold Burtt reported that the systolic blood pressure method could correctly detect experimental crimes in "91 per cent of the cases as compared with 73 per cent" using the breathing method.[33] Giving no experimental details, Keeler reported a response time–word association study that detected deception in sixty-two percent of cases. However, blood pressure and respiration techniques improved accuracy to ninety-three percent.[34] Fred Inbau maintained that experimental cases were accurate "approximately eighty-five per cent" of the time, adding that in criminal cases, full confessions had been obtained in "approximately seventy-five per cent of those in which the record indicated deception."[35] In his history of lie detection, Paul Trovillo reported that the Scientific Crime Detection Laboratory examined 2,171 subjects between 1935 and 1938. Of this number twelve mistakes in diagnosis had been verified, wrote Trovillo, who calculated that the errors were of the order of "five-hundredths of one per cent." And even if the errors were ten times this number, "they would still be a relatively small proportion of the total."[36]

Not everyone let the statistics go unexamined. In 1939 Walter Summers criticized Marston's work "because of its impressionistic character, so that the apparent statistical result is valueless as a critique of the accuracy of the procedure." Having quoted accuracy claims by Larson and Keeler, Summers concluded that a procedure that started with an experimental validity of only eighty-five percent was "an extremely hazardous thing to employ in the investigation of the guilt or innocence of any person." "Even the 75% efficiency obtained in the numerous criminal cases leaves a very great probability of error.... The 75% efficiency by no means tells us the entire story, for it fails to relate the number of instances in which deception was actually practiced in a manner which eluded the examiner and the instrument." Despite his forceful criticism, Summers did not dispense with the accuracy statistic altogether. The preliminary results obtained with his own psychogalvanometer—an instrument rejected by both Marston and Larson—had "showed an efficiency of better than 99%."[37] "Caution Keeler whenever you see him to cut out his talk about the infallibility," Larson asked Vollmer in 1931, "because I know it is not infallible. Tell him to never have it used so that men are discharged because of the interpretation of the record, or legal action of any sort taken."[38]

The status of invention and the power of numbers were not the only ways advocates of instrument gained legitimacy. Another one was the depiction of the machine as a "black box." Expository articles often included a photograph of the enigmatic instrument, a depiction that explained yet mystified at the same time. Here was a gadget fabricated from reassuringly complex components, all of which were encased within a scientific-looking "black box." By describing the instrument thus, however, the question what exactly it was fabricated from remained unanswered. The first explanation conveniently rendered the second superfluous: that the instrument looked "scientific" was a sufficient testimony of its credibility. What exactly was within the black box was rarely explained. It was useful to have the lie detector described as "mysterious," or as the *Saturday Evening Post* put it, as a "curious engine."[39] Henry Pringle began his 1936 *Reader's Digest* article "How 'Good' is Any Lie?" with an account of the experience of submitting to a lie detector test: "We sat in a small room at the Scientific Crime Detection Laboratory of Northwestern University in Chicago. On a table behind me rested a small, box-like machine. This was the Keeler Polygraph, popularly known as the 'lie-detector,' and I was being subjected to an examination on my truthfulness. Professor Leonarde Keeler of the Northwestern University law school, was conducting the test." Having praised the abilities of the intimidating "Keeler Polygraph," Pringle wryly observed that the "vast majority of defendants . . . are entirely confident that they can outwit the little black box."[40] Keeler described his machine in the same manner. In 1948, he told Vollmer that a fellow operator had just completed his six-week polygraph course. "He is now carrying one of our black boxes to the Orient," he wrote, "where he will be stationed for some time."[41]

Reporting on a sensational "eleventh-hour" lie test that "sealed the doom" of a convicted murderer, the *Literary Digest* described how Keeler "put an odd looking black boxlike machine, about two feet square, on the table" in the jail cell.[42] In 1936, the *New York Times* described how a school principal wrecked his "experiment in psychology"—a homemade lie detector—that was attracting media attention.[43] "The destruction of the black box equipped with dials and electric bulbs closed the incident." In 1937, *Scientific American* pictured an instrument replete with dials, switches, a graphical recording device, and an assortment of tubes.[44] The components looked scientific and complex, but the technical language obscured the fact that they were quite straightforward. Describing "Two Simple Ways To Make A Lie Detector," *Popular Science Monthly* hit the nail squarely on the head when it told readers that "since the device is so simple, it is advisable to conceal it in a box so as to hide the mechanism and give it as mysterious and complicated an appearance as possible." Although the homemade machine was little more than a toy, it nevertheless worked "on the same principle as the famous lie-detecting instruments used by criminologists in obtaining confessions of guilt from law violators."[45]

The semblance of science served not only to legitimate but also to threaten. The black box was not only mysterious but also frightening. Although the lie detector's apparent role was to replace the third degree, it never managed to lose its intimidating character. This feature of the test was recognized in 1929 when the *Literary Digest* described the process when a suspect faces "a new kind of third degree, a strange machine that neatly separates falsehood from truth in the story he tells."[46] The *Science News Letter* might have been somewhat cynical, but its point was well made: "Chief usefulness of the gadget is as an aid to the police in scaring an ignorant or superstitious person into making a confession of crime. An empty black-box, if it looks mysterious, would serve the same purpose—and has been used for it."[47] In fact, an "empty black-box" had already been used to induce confessions. In July 1931, the Philadelphia Police obtained a confession of guilt from a boy with a lie detector made up of stolen radio parts. During the test, which was a curious mixture of modern science and ancient magic, the young suspect was seated in front of the contraption and told that if he told a lie "it would be reflected in a color register behind his back. . . . Wires were placed on his arms and legs, the police told him, to register his 'blood flow' in the device. All lights save a small red one were extinguished. Each time the police thought the boy's answers to questions were wrong they told him the 'lie detector' showed green. Then pepper seeds were put on his tongue. He complained that they burned. 'Those pills always burn the tongue of a liar,' he was told." After the youth began to cry he made a statement implicating himself and five others in a number of recent robberies.[48] The "lie box," as one suspect described it, was extremely intimidating by virtue of its apparent infallibility.[49]

The rhetoric of the machine's scientificity was completed by an image that would come to achieve iconic status. As the graphical output of the black box, "the chart," achieved immense persuasive power as the essential record of truth or lies. The first illustration of such a chart appears to have been published in Balmer and MacHarg's Luther Trant (1910) story "The Man Higher Up." In the story "Professor Schmalz" uses a plethysmograph and a pneumograph to detect a person's disliking of caviar: "The instruments show that at the unpleasant taste you breathe less freely—not so deep. Your finger, as under strong sensation or emotion, grows smaller, and your pulse beats more rapidly."[50] In 1932, *The Review of Reviews* provided what would become the standard image. Accompanying a photograph of Keeler performing a lie detector test were pictures of two short strips of graph paper. The top graph, showing a gently undulating curve, was captioned "THE TRUTH." Below it, another graph described a violently fluctuating line and was captioned "THE LIE."[51] A similar "Lie and Truth" image was chosen by *The Literary Digest* to illustrate an article, "Detecting Liars."[52] "One subject made both records," read the caption. "The upper is the 'lie' record, the lower one is when he told the truth." In a later piece, "Catching Criminals with Lie-Detector," the magazine printed a photograph of "The blood-pressure record of an embezzler who decided to go straight." The upper section, "taken before he confessed, shows characteristic tension, caused by lying. The lower section, after full confession, shows steady, normal pressure."[53] Such comparison graphs were very popular.

Sometimes only a single strip of graph paper was shown. In such cases, the graph would invariably undulate smoothly until interrupted by a disturbance that signified the lie. An arrow might point the reader to the crucial moment. The legend on *Scientific Monthly*'s illustration was "The arrow on each record indicates the peak of tension in the subject's blood pressure curve—the point at which the lie was told."[54] "A black arrow points to the jagged peak," a 1937 *Literary Digest* illustration was captioned, "depicting a lie which sent Joseph Rappaport to the chair for murder."[55] The lie occurs at an apparently simple and discrete moment, and that moment could be detected, and its parameters calculated. Readers were led to believe that the graph paper could almost speak for itself, so obvious was the appearance of the lie. According to the rhetoric of the image, the chart did not need a human operator because the machinery seemed to work so well as to not require one: the truth was plain for all to see. Images of the chart obscured the operator, whose role it was to scrutinize the chart in order to render it intelligible. *Reader's Digest*'s reporter was glad that the questions of his mock examination had not been too embarrassing, "for the wig-wag lines on the polygraph's recording roll of graph paper indicated all too clearly when I had lied."[56] Responsibility for determination of truth was transferred from human to machine: "The needles flickered unsteadily, indicating, Professor Keeler said later, that Rappaport was lying. . . . Professor Keeler turned to casual questioning. The needles graphed a steady course."[57] Because the physiological detection of deception was essentially an interpretive human enterprise, an edifice had to be constructed around the instrument to deflect criticisms of its subjective and perhaps arbitrary nature. Although human scrutiny was necessary to interpret the curves on the chart, discourse about the lie detector worked to hide this fact.

Yet, paradoxically, the lie detector's advocates were anxious to stress that their expertise was indispensable for the success of the venture. Such a tension was part of the larger problem of the presence of the polygraph operator. If a machine could detect a lie automatically, why was an expert required? The operator was potentially a mere technician, a cog in the machine. Yet he was also the expert, the sage, and, ideally, also the instrument's inventor. The charismatic authority of the expert mediated this fine balance between these two opposing poles.

Not everyone found the rhetoric persuasive, however. In 1939, A. A. Lewis criticized the myth of the autonomous black box in an article for the *Scientific Monthly*. "Diogenes is back again with his lantern," he wrote. "But this time the lantern looks too much like a laboratory to be regarded as a trick of burlesque employed to incite an insurrection against dishonesty": "These detective instruments, like an X-ray machine, may turn culprit or criminal inside out, as by zigzagging curves and dial readings his deeper bodily changes are made visible, but it still remains to interpret the picture. The suspect's guilt or innocence is not spelled out in unmistakable letters. The fact that these deeper, organic reactions are involuntary and can't be made to belie culpability like a face 'kept straight,' does not guarantee that they can be subject to no other source of causation except actual misconduct." By calling the lie detector operator a "laboratory magician," Lewis hoped he would undermine the claim made by advocates that the instrument had scientific credibility.[58] In fact, attributions of magic actually supported the agenda of lie detector advocates.

The scientific "chart" so easily became a numinous "scroll." Sacred truths were being revealed. An illustration of a "typical graph" in *Scientific American* encapsulated the conjunction of science and magic: "This graph depicts the sudden rise in blood pressure at the point of attempted deception. The subject was handed ten well-shuffled playing cards, with instructions to choose a card and then lie about his choice. Respiration at top, blood pressure below. Notice where he said 'No' to the three of diamonds. He later admitted that the three of diamonds had been his choice."[59] Observers were led to believe that lie detector could perform card tricks. Although the ostensible purpose of the "stim test" (as it would become known) was to "obtain controls," the trick also functioned as a remarkable demonstration of the machine's abilities. "Control readings" were normally obtained through the use of innocuous or irrelevant questions. But the importance of the card trick lay in its power of intimidation.

Consider, for example, the point made by Father Summers, inventor of the psycho-galvanometer lie detector, to a reporter for *Forum and Century* magazine. He was shown ten playing cards and asked to make a mental note of one of them, "keeping my selection a secret. . . . I chose the deuce of hearts; the cards were shuffled and then displayed to me one at a time. 'Is this your card?' asked Father Summers, as each was turned up. I steadfastly replied, 'No,' keeping my eyes on the galvanometer dial to see what happened. When the deuce of hearts appeared, I said, 'No,' as coolly and disinterestedly as possible, but the indicator shot up like a jack-in-the-box. After two repetitions of the test, I broke down and 'confessed' that I was lying about the deuce of hearts." "So you see," said Father Summers, "if one perspires over a little thing like a playing card, what would happen if a real crime were being concealed."[60] The trick was already a component of the orthodox examination by the early thirties. "So delicate is the apparatus," reported the *New York Times* in 1931, "that a subject can be asked to select a card from a deck and will react at once when the correct card is picked up."[61] In 1937, Keeler used the technique on Joseph Rappaport, the convicted murderer whose eleventh-hour jail cell lie test sealed his fate.[62] A photograph accompanying a *Newsweek* story on the lie detector showed Keeler revealing a playing card to a female subject strapped to the lie detector. "Professor Keeler's card trick works nine times out of ten," read the caption.[63] The entire procedure was a classic piece of misdirection. What looked like the work of the instrument was actually a ruse engineered by the operator using a marked playing card. Keeler, a keen amateur magician, had devised the sleight of hand.[64]

The lie detector was considered magic because it embodied scientific progress: it discovered the truth, promoted justice, and was humane. "The machine has now been used in 60,000 cases," the *Saturday Evening Post* enthused, "and its uncanny power of penetrating guilty secrets has been thoroughly established." The lie detector was awe-inspiring because it was thought that it could achieve its impressive results on its own, without human assistance: "Automatically controlled pens will record the slightest deviation from the truth." The attribution of consciousness was irresistible: "When a lawbreaker denies his crime during a lie-detector examination, the pens become feverishly animated. A guilty man, seeing that the machine is practically photographing his soul, usually cuts short the examination by confessing."[65] What was most magical about the lie detector was its "uncanny power" of agency. It was thought that the machine, not the human operator, did the work. The *Saturday Evening Post* articles were peppered with personifications and attributions of agency: "The lie detector acts as a mechanical conscience"; "the machine had solved some baffling mysteries"; "The detector has a peculiar genius for geography"; "It can read a nervous, excitable person like a book; it can read a tough, hard-boiled character like a book"; "It airs a scandal in a sorority house, stops students from cribbing, bank presidents from tapping the till, and releases a guiltless man condemned to a lifetime in jail."

The instrument was often described as possessing human body parts, such as "an accusing finger." In its review of electronic devices used for "the detection and prevention of crime," *Radio News* asserted that the field was "richly aided by the sharp eye and keen ear of electronic devices of one sort or another."[66] "Three moving fingers of the Keeler Polygraph record these changes," observed *Reader's Digest*.[67] And if it possessed a body, then it also possessed a mind. "The vast majority of defendants," claimed the magazine, "are entirely confident that they can outwit the little black box." A subject experiencing a lie test "made vigorous denials, but the polygraph betrayed him." The instrument habitually took on a persona in these narratives, such as the street-wise detective. *Reader's Digest* introduced its piece with the caption "The 'lie-detector' at work solving crimes."[68] "To the lie detector goes the credit for 'cracking' this strange enigma," remarked *Scientific American* about one particularly puzzling case, "and searching out the murderer from among nearly 50 suspects, as well as locating the murder weapon." It was a "machine that knows all the answers," promising to show how "Leonarde Keeler's astounding invention tracks down murderers, unmasks the liars and the larcenous, and can tell you just how honest you are—and intend to be."[69]

A. A. Lewis described the instrument as a "mind reading device."[70] Although the lie detector's scientific credentials apparently contradicted its supernatural abilities, both qualities were acceptable within a broader context that encouraged the responses of awe and intimidation.

"It is difficult for an accused person to escape taking the test," wrote *Reader's Digest*; "to refuse is a fairly clear admission of guilt."[71] Because the test was depicted as being a scientific and humane interview technique, it became virtually impossible to refuse to take it without incurring a suspicion of guilt. Three "ace" detectives were demoted to patrol duty by the police committee for their unwillingness to submit to lie detector tests during a District Attorney's investigation of gambling, reported the *New York Times* in 1938.[72] "Ordinarily, a suspect cannot be legally compelled to face the lie detector," said the *Saturday Evening Post*. "It doesn't look good, however, for one who proclaims his innocence to refuse. Nearly every accused man pretends to welcome the test."[73]

If a refusal to take the test signified guilt, volunteering to take the test signaled innocence. It was often worth the gamble because a suspect could be completely exonerated by a successful lie detector test. In 1935, for example, a lie detector test virtually cleared a Fairfield carpenter of suspicion of the murder of a young girl.[74] Commenting on a different case a month later, the newspaper told its readers, "The lie detector said [the suspect] was innocent. He was released."[75] In 1933 a convicted inmate was freed from the Marquette Penitentiary in Michigan for having passed a lie detector test conducted by Keeler.[76] "Lie-Detector Test Asked by Prisoner" read a *New York Times* headline in 1935: "Paroled Convict Would Have Machine Prove He Did Not Steal Automobile."[77] The same year, the accused Lindbergh baby kidnapper Bruno Hauptmann requested the opportunity to take a "truth test." He also asked that an important prosecution witness take one too.[78] Hauptmann understood the symbolic power of passing the test. The authorities knew it too, which might explain why his request was disallowed even though Marston had offered to conduct it himself.[79] In 1936, Jackie Coogan and Betty Grable, "dancing sweethearts of the films," volunteered for polygraph tests in order to prove that their story about a robbery was true.[80] Requested to take a test to prove his innocence of fixing a prize fight, boxing promoter Nick Londes called the bluff on the police: "'I'll take a lie test if the Police Commissioner will,' he snorted."[81] Anyone who volunteered to submit to the "lie box" was assumed innocent. Similarly, anyone who refused a test was considered more than likely guilty. Being prepared to take a test was thus akin to being prepared to take one's case to court. Although it clearly represented the processes of law and order, it was as though the lie detector functioned as a legitimate alternative to the justice system, at least in public discourse.

The concept of the lie detector serving as an alternative legal system was further supported by its reputation as a test that could not be beaten. As expository articles made clear, polygraph records were either "innocent" or "guilty"; there were no intermediate possibilities (save perhaps "inconclusive," which necessitated a retest).[82] Like the law itself, the test was apparently applicable to all. But certain types of persons could beat the test. "Certain psychopathic subjects show abnormal irrelevant responses," Marston wrote. "A feeble-minded person may not comprehend the situation sufficiently to become conscious of deception or its implications."[83] "Morons and psychopaths either don't react or react wildly," the *Saturday Evening Post* confirmed, "so that it is often impossible to tell whether they are lying or not."[84] Only the trained expert could beat the machine, reported the *New York Times*. "Keeler himself has admitted that he can fool his own instrument," it wrote, "but that is because he is so familiar with its working."[85] In principle then, the instrument could be applied to everyone except those such as "the moron," "the psychopath," "the feeble-minded person," the child, and the lie detector expert. Beating the machine, like beating the legal system, was considered impossible for the vast majority of ordinary citizens. The polygraph was powerless against the pathological however. Unlike instruments such as the "soul machine," whose advocates a generation before had deliberately sought out "morons" and "psychopaths" to study, the lie detector only worked on normal people.

The lie detector as an alternative legal system was most saliently evidenced by a set of metaphors centered on the law court. "The innocent man finds in the machine his most reliable witness," said the *Review of Reviews*.[86] "It was the fourth time the lie detector has been admitted as a witness," said the *New York Times*.[87] Reporting on the Rappaport case, the newspaper wrote that the "verdict of the recording needle was: 'He lies!' "[88] The *Literary Digest* wrote that the lie detector "had sealed his doom by returning the same verdict as the human jury of his peers: 'guilty.' "[89] The instrument "helped to bring another slayer to justice," wrote the *New York Times* in 1935, as a suspect was betrayed "by the scientific crime revealer."[90] Later that same year, the newspaper reported on how use of the instrument exonerated a suspected burglar: "The lie detector said he was innocent. He was released."[91] "The lie detector gave

Goldman 100%," reported *Time* magazine in 1944, "and Judge Leibowitz gave him his freedom."[92]

The lie detector, asserted *Outlook and Independent* in 1929, "seems little more than another method of wringing out a confession after protracted questioning."[93] By describing it in the same terms as "a piece of rubber hose" and "a makeshift electric chair," the magazine was ridiculing the claim that the instrument exhibited scientific humanitarianism. Such a claim was diametrically opposed to those of the instrument's advocates who unanimously presented it as an antidote to the notorious third degree. The *Outlook and Independent* recognized that the contrast between progressive science and reactionary brutalities was not so great as it appeared. Intimidation was an important quality attributed to the machine. That the lie detector resembled an electric chair was, therefore, not a particularly troubling problem. *Scientific American*'s description of an examination of a "Negro porter" deployed a mixture of scientific rhetoric and intimidation: "The whites of his large round eyes were an extreme contrast to his face, and he shied at the boxlike instrument sitting on the table. After much explanation and assurance that the instrument would not injure him, the porter was finally induced to sit in the chair and allow the pneumograph to be adjusted about his chest and the blood-pressure cuff attached to his arm."[94] The porter's anxiety was well founded because of the common belief that the lie detector delivered an electric shock to the subject. To the uninitiated, the lie detector looked very much like a technology of execution. Both instruments were symbolically organized around the theme of the chair, the seat of final judgment. *The Nation* deliberately undermined claims that the instrument embodied humanity by comparing it to an electric chair.[95] From the gathering of evidence to the carrying out of the sentence, the lie detector was considered competent in every step in the "fight against crime." " 'Lie detection' by dependable scientific means," recognized *Living Age* in 1935, was "the constant dream of jurists and police functionaries."[96] And because "every crime was entrenched behind a lie,"[97] the lie detector therefore embodied nothing less than the dream of criminology.

As criminal anthropology had done so before it, so the lie detector technique also accorded women a privileged place in its moral economy. Female bodies were depicted as setting a unique challenge to the lie detector. Emotions attributed to women—love in particular—played an important role. Both Larson and Keeler met their wives through the instrument, and William Marston credited his with the discovery of his main principle: "I shall always

By the late 1930s, the "male gaze" had been entrenched within orthodox lie detector procedures. Paul V. Trovillo, "A History of Lie Detection," pts. 1 and 2, *Journal of Criminal Law and Criminology* 29 (1939): 848–81; 30 (1939): 104–19. Reprinted by special permission of Northwestern University School of Law, *The Journal of Criminal Law and Criminology*.

be grateful to 'the girl from Mount Holyoke' for suggesting the idea that deception makes the pulse beat harder, and for assisting throughout the original research which established the systolic blood pressure deception test."[98] In the Luther Trant story "The Man in the Room" (1909), a woman, surrounded by four men, is seen lying in bed and speaking into a mouthpiece as part of a reaction time–word association test: "'Dress!' he enunciated clearly. The pendulum, released by the magnet, started to swing. The pointer swung beside it in an arc along the scale. 'Skirt!' Miss Lawrie answered, feebly, into the drum at her lips."[99] One notable feature of Trovillo's paper was a photograph of an ideal lie detector test situation.[100] Taken from an elevated position so the viewer could observe the arrangement of the test, the photograph showed a man and a woman sitting on opposite sides of a desk, between them a Keeler Polygraph. The woman sits parallel to the length of the desk, gazing ahead into the middle distance. Her right arm rests on the desk, her left on the arm of her own chair. A blood pressure cuff is attached to her right arm, a galvanometer electrode to her left hand, and a pneumographic tube has been wrapped around her chest. The smartly-dressed male examiner sits on the other side of the desk, his right hand holding a pen poised to write. He is staring intensely at the woman.

The demarcation between examiners and subjects was evident in the earliest visual depictions of the examination. In the illustration accompanying the *Boston Sunday Advertiser*'s 1921 piece on Marston's apparatus, the "suspect" faces the "Questioner" across a table while a second man takes the suspect's blood pressure.[101] A similar arrangement can be seen in the photograph featured in the 1924 *Collier's* article. Keeler fiddles with the apparatus behind the subject's back while the examiner faces the suspect and holds the chart in his hands.[102] A photograph in a 1929 magazine article shows a subject attached to the machine facing his interrogator, a policeman, while another policeman and two surly looking detectives look on.[103] The frontispiece to Larson's *Lying and Its Detection* shows a policeman giving a lie test to a disheveled-looking male subject.[104] Two other photographs in Larson's book show subject and examiner surrounded by numerous other observers. The early photographs tended to feature male subjects, male police examiners, and any number of observers. A 1934 magazine photograph, for example, shows Leonarde Keeler, Calvin Goddard, and a third man observing a seated male subject while a police officer looks on.[105]

During the 1930s, however, images of lie detector examinations were gradually standardized. Keeler is seen sitting on the desk behind a male subject in one picture, standing behind him in another.[106] A 1935 newspaper photograph shows him standing behind a female subject while she gazes passively ahead.[107] *Popular Science Monthly*'s photograph was also of a female subject and a male examiner.[108] By the late 1930s, the archetypal image was of a male examiner and a female subject, as in the Trovillo article. In one picture, Marston can be seen explaining the results of a lie detector test to a beaming young woman still strapped to the instrument.[109] Alva Johnston's 1944 series of articles opened with the classic image. Seated to the rear of the photograph, Keeler operates his desk polygraph and gazes at the female subject who is looking ahead into space.[110] A 1951 photograph features a standing male watching a seated female. He is wearing a white coat.[111] The image of a male examiner actively gazing upon a passive female subject has since become a standard feature of lie detector test images.[112]

Not all visual portrayals of lie detector examinations feature female subjects. Nevertheless, considering that male offenders vastly outnumber female offenders, women are certainly disproportionately represented. Female subjects tend to be used in ideal examination scenes such as those found in textbooks. The first photograph in Reid and Inbau's *Truth and Deception*—widely regarded as the essential polygraphy text—features a male examiner gazing upon a female subject.[113] Other expository texts also use the male expert/female subject.[114] Trovillo, Keeler, Reid and Inbau, and Matté all used women in their posed photographs of the examination situation.[115] Why were women "ideal" polygraph subjects? Not surprisingly, Marston had something to say on the subject. "Women, agree masculine sages," he wrote in 1938, "are the worst liars. But are they?" "Treatises have been written—by men—to prove that women lie more frequently because they are the weaker sex and must deceive continually to protect themselves": "Women earn their livings mostly by deception, some cynics assert, pretending affection for men they don't love and tricking men they do love into unwilling generosity. But that sort of arm-chair indictment of the fair sex's truthfulness need no longer go unchallenged. The Lie Detector now supplies a method for scientific comparison between male and female truthfulness." Not wishing to refute the apparent wisdom of gender differences in honesty, Marston merely wanted to replace supposition with truth. He concluded that "men are more dishonest in business and women in society." Women apparently told "innumerable lies . . . to enliven social conversations and to manipulate other people for various petty purposes or oftentimes just for the fun of it."[116]

Marston's gender dimorphism was not unusual. In a 1938 feature, "A Machine to Measure Lies," *Look* magazine reported Dr. Orlando F. Scott's belief that "women respond with so much electrical energy that their lies are easier to detect than those of men."[117] "It Really Understands Women" read the caption to a newspaper photograph of a woman being given a "photopolygraph" examination,: "All Emotional Reactions Recorded."[118]

The iconic image of male examiner and female subject permits contradictory readings, however. As Marston and his contemporaries suggested, women were either "more emotional" than men or their honesty varied with the situation. In either case, they were shown as inferior to their male examiners, whose scientific authority positioned them as objective, truthful, and composed. The female subject, however, can perhaps be read as representing an authentic conjunction of truth and nature. Women were putatively more "naturally emotional" than men and thus better subjects for polygraph examination. The notion of nature as a woman unveiled by science has a long history stretching back to the scientific revolution.[119] In this model, masculine science is a form of power wielded over feminine "nature." In pictures of lie detector tests, examiners are invariably clothed in the vestments of authority, such as the police uniform, the scientific white coat, or the business suit. Female subjects, in contrast, are usually casually attired. In some photographs the tight-fitting pneumographic tube accentuates the subject's breasts.[120]

The erotic meaning inherent in the scientific "lifting of the veil" is echoed in the ideal polygraph examination scene. Male examiners were shown gazing intensely at their female subjects, ostensibly seeking "behavior symptoms" of deception.[121] Examination rooms are designed to provide numerous opportunities for voyeuristic surveillance via two-way mirrors and concealed microphones.[122] The division of gender and power extended to the instrument itself. Lie detectors were evidently "toys for the boys"; the technique rendered masculine by virtue of its association with the heroics of invention and crime fighting.[123] None of the early pioneers were women, and there are relatively few female polygraphists today. In 1975 the *Journal of Polygraph Science* thought the story of a female examiner sufficiently notable to warrant the front-page headline, "Why a Female Polygraphist?"[124] Female examiners were reputedly more successful than their male counterparts on account of the liar's "mother-complex," an inability to lie to a "mother-figure."

According to Roland Barthes, the function of myth is to express and justify the dominant values of a given historical period. Barthes claimed that "The Brain of Einstein," for example, was a "machine of genius," symbolizing the power of thought and embodying "the most contradictory dreams." Einstein was a machine and a magician, a tireless researcher and a romantic discoverer.[125] Barthes concluded that myth is unconcerned with contradiction so long as it establishes a "euphoric security": a credible and intense ideological edifice. Lie detection was such a mythic and ambiguous enterprise, a manifestation of contradictory notions such as science and magic, freedom and coercion. The various signs circulating through the discourse were not isolated but were invariably linked in incongruous pairings. The black box was scientific but also intimidating. The scientific instrument could perform magic. The humane technology that promised to replace the third degree was threatening. While it functioned automatically, the lie detector nevertheless apparently possessed sentient agency. Thus, although the discourse of lie detection was loosely homogenous, it was far from being internally consistent. Such contradictions were unimportant, however; what mattered was that the complex of signs rendered the lie detector practically workable, socially acceptable, and culturally meaningful. From about 1921, the lie detector became an important resource for newspaper reporters and magazine writers wishing to find a symbol of a new approach to crime fighting. To a considerable degree, the lie detector was created in those pages. Lie detector discourse emerged primarily in publications like *Collier's*, the *New York Times*, the *Popular Science Monthly*, and the *Saturday Evening Post*. These were the vehicles that launched the lie detector on its quest for euphoric security.

CHAPTER 8

"A bally hoo side show at the fair"
The Spectacular Power of Expertise

> What are the facts about razor-blade quality? That's what Gillette wanted to know. And that's why Gillette retained Dr. William Moulton Marston, eminent psychologist and originator of the famous Lie Detector, to conduct scientific tests that reveal the whole truth. Truck drivers, bank presidents . . . men in every walk of life . . . take part in this investigation. Strapped to the Lie Detector . . . the same instrument used by police . . . these men shave while every reaction is measured and recorded.
> —*Life* magazine advertisement (1938)

> Looking back on that period, it seems to me now that while Nard had carried on hundreds of experiments with the lie detector and had worked on numerous police cases during his college years, he was still waiting in the wings. The stage was set, the actors ready, but the curtain had not yet risen on the big show.   —E. Keeler, *The Lie Detector Man* (1984)

On August 31, 1937, *Look* magazine explained "How a Lie Detector Works."[1] The article was illustrated with a photograph of a subject being examined and diagrams of two sections of polygraph chart. One was captioned "THE MAN IS LYING," the other "THE SAME MAN TELLS THE TRUTH." "If you tell a lie, you upset your emotions," the piece clarified. "Your breathing becomes heavy, your blood pressure increases." "That's why Leonarde Keeler of Northwestern University's crime detection laboratory has been able to invent a machine which, he claims, can test the truth of statements. The Keeler Polygraph or lie detector is used to determine the guilt or innocence of crime suspects. Many thousands of such tests are made monthly all over the U.S." The following year, on December 6, 1938, *Look* magazine published another feature about the lie detector. This time, Marston was credited with inventing the instrument. "Would YOU Dare Take These Tests?" was the headline to the double-page photo story: "Real Life Stories from a Psychologist's Files." Originating from "the Field of Crime," the lie detector had now entered "the Fields of Love." It was capable of telling you whether or not your wife or sweetheart loved you, and vice versa: "Dr. William Moulton Marston, the inventor, reports success with his device in solving marital or other domestic problems and adds that it will disclose subconscious secrets of which the subject is utterly unaware."[2] Not only did the machine discover that "the neglected wife and her roving husband" still harbored some affection for each other, but it also revealed that a young couple were in love, despite being engaged to other people. Once the "disinterested truth-finder" had diagnosed the cause of the symptoms, the consulting psychologist was able to confer his blessings on the unions. "United by the lie detector," the happy couple thanked Marston for recommending marriage.

These two magazine articles reveal much about Marston and Keeler and their respective agendas for the lie detector. For Keeler, of Northwestern University's crime detection laboratory, the instrument was an important weapon in the fight against crime. For Marston, a famous consulting psychologist, it could be used to resolve marital or other relationship problems by disclosing secrets unknown to the subject. Reporting on Keeler's infallible lie detector back in 1924, *Collier's* had reassured its readers that there was "no immediate danger of the lie detector following the talking machine and the radio set into the intimacy of domestic life."[3] By the late 1930s, this is exactly what Marston was enthusiastically advocating.

Marston received an LLB from Harvard Law School in 1918 and a PhD in psychology three years later.[4] His doctoral research was concerned with the physiological detection of deception. Having performed experiments with the so-called "systolic blood pressure deception test," and after testing it on suspected spies during the First World War, in 1923 Marston unsuccessfully attempted to get his deception test admitted as evidence in a court of law.[5] During the early 1920s Marston devoted himself to empirical research on the detection of deception and the measurement of systolic blood pressure. In 1924 he traveled to New York City to work with the National Committee for Mental Hygiene, and then to Texas, where he later claimed to have analyzed "every prisoner, male and female, in the state penitentiaries" according to his theory of the emotions.[6] He later advanced a "psychonic theory of consciousness," debated the relative merits of materialism and vitalism, and speculated on the relationship between "primary colours and primary emotions."[7] This productive period culminated in his first book-length study, *Emotions of Normal People* (1928).

In 1928 Marston used his systolic blood pressure deception test to investigate the emotional responses of "blondes, brunettes and red-heads." The experiment was reported by the *New York Times*, which was enthusiastic if skeptical. "Blondes Lose Out in Film Love Test," it proclaimed. "Brunettes Far More Emotional. Psychologist Proves by Charts and Graphs. Theatre a Laboratory": "By elaborate and allegedly delicate instruments known in scientific circles as the sphygmomanometer and the pneumograph, by charts and graphs, and by the simpler expedient of holding hands, Dr. William Marston, a lecturer on psychology at Columbia University, proved yesterday in the presence of a staff of coy press agents, camera men, motion picture operators and columnists that brunettes react far more violently to amatory stimuli than blondes."[8] The Embassy Theatre was an appropriate setting for the vaudevillian experiments. The technique involved strapping women to the apparatus and showing them clips from movies such as the Greta Garbo–John Gilbert pictures *Flesh and the Devil* (1926) and *Love* (1927). "The experiments more or less proved," said the *New York Times*, dutifully reproducing Marston's interpretation of events, "that brunettes enjoyed the thrill of pursuit, while blondes preferred the more passive enjoyment of being kissed."

Marston was preoccupied by popular psychology from the mid-1930s. "His versatility enabled him to break into the high-class magazine field at once," a biographical piece immodestly recalled, "and he has written articles for all the leading magazines, besides many newspaper articles and popular books."[9] A 1934 *Chicago American* magazine article—by "Prof. WM. M. Marston (Famous 'Practical Psychologist')"—was typical. Its inelegant title was "Science Derides the 'Love-Slave' Verdict, Crying '*Woman* is the *Man's* Love-Master.'"[10] The piece was a discussion of a recent New York City murder trial. Defendant Marquita Lopez claimed "she had acted under the compulsion of a man to whom she was passionately devoted; that as a 'love-slave,' she had merely done what her love-master desired." She was found not guilty. Marston's academic work had well prepared him to analyze true crime in a sensational manner, and his favorite psychological categories of dominance, submission, inducement, and compliance were perfectly suited to discussing the Lopez case. Psychological experiments, he claimed, had proved that "men really prefer to *submit* in the love situation, while women prefer to *induce*."[11]

The Spectacular Power of Expertise   157

In 1928 Marston used his systolic blood pressure deception test to investigate the emotional responses of "blondes, brunettes and red-heads." Frontispiece to William Moulton Marston, *Integrative Psychology: A Study of Unit Response* (New York: Harcourt, Brace and Co., 1931).

"To excuse her from criminal liability for murder on this ground," Marston wrote, "is as psychologically antiquated as it is to believe that man is the master in a love situation." Although circumstances forced him into leaving the academy to forge a career as a "consulting psychologist," Marston discovered that his theories had wide application in the public domain. Unusual as they were, his ideas allowed him to become widely known.

Although Marston made a transition from professional academician in the 1920s to earnest populist in the 1930s, he maintained his faith in psychology as a liberating force for good throughout his life. It was a vision he might have acquired from his mentor. Hugo Münsterberg had not been averse to courting public controversy by involving himself in social and political causes. Nor had he been dissuaded from writing mass circulation popular psychology articles by condescending colleagues who dismissed them as "yellow psychology."[12] Münsterberg had a broad spectrum of philosophical and psychological interests, but it was the psychology of deception that his undergraduate student found most fascinating. Many years after Münsterberg's death, and in a dubious tribute, his student recalled the Oedipal moment when he confronted his mentor with his new theory of deception: "I had been working on the Jung reaction-time test, I remember, and I was in despair," Marston recalled. But he had a "half-baked idea" that would nevertheless "mean a new theory of deception": " 'I've watched my subjects carefully. When they lie they seem to put more effort, more dominance or self-assertion into their story. That increased effort which, if I am right, is called forth by the act of lying, ought to make the heart beat harder. And if the heart beats more strongly then the blood pressure must go up.' " Münsterberg apparently dismissed the idea at first, saying it was very theoretical.[13]

In 1941 Marston created *Wonder Woman*, "a feminine character with all the strength of a Superman plus all the allure of a good and beautiful woman."[14] He provided her with a lie detector of her own—a "Golden Lasso of Truth." The character first appeared in the November 1941 issue of *All-Star Comics*.[15] The following summer she starred in her own title, and by the end of the year was appearing in four different comics.[16] She was immensely popular. By the third issue alone, the comic book was selling half a million copies.[17] As with his popular psychology, so with his super heroine, Marston structured Wonder Woman's moral universe with the categories of dominance and submission. She was constantly being chained, imprisoned, tied up, handcuffed, and blindfolded.[18] While such plot devices allowed her author to construct entertaining situations to challenge her ability and ingenuity, they also rendered visible his deeply-held philosophy of freedom. It was a philosophy that had initially originated during his early work with the lie detector.[19]

By the late 1930s, Marston had become a well-known public figure. Two factors were responsible for this. First, his theoretical ideas about the "primary emotions" of dominance and submission were perfect for discussing sex.[20] Dominance and submission were loaded with sexual meaning, but they were also sufficiently flexible to be used to interpret a wide variety of social and political situations, such as crime or military aggression. Armed with such a philosophy, he found making the transition from academician to populist relatively straightforward. That he had an extroverted personality and was adept at dealing with the media also helped. Second, Marston was an enthusiastic advocate of the lie detector. In 1938, for example, he appeared in a series of advertisements in *Life* magazine for Gillette razor blades with the instrument. "Lie Detector 'Tells All,' " announced the advertisement's mock headline, "Reveals Startling Facts About Razor Blades!"[21] The accompanying photograph showed Marston reading a lie detector chart from one of three men who were busy shaving. Non-Gillette blades evidently produced "emotional disturbances" in the subjects: "9 out of 10 men tested by Mr. Marston express preference for Gillette blades."[22]

Marston refrained from publishing a book about the lie detector until 1938, even though he claimed to have discovered the principle upon which it was based in 1915. There was no mention of the lie detector or deception tests in his major work *Integrative Psychology*, and his *Emotions of Normal People* only discussed the deception tests in the context of "Abnormal Emotions." In fact, his academic work had virtually ignored the lie detector. By the late 1930s, however, Marston had become a public personality through his newspaper and magazine columns, popular psychology books, and radio appearances. A captivating character, by 1938 he had successfully established himself as a psychologist in the public sphere. Only when Marston became a writer of popular psychology did the lie detector come to play a greater role in his career.

The Lie Detector Test (1938) was less a professional training manual (as John Larson's 1932 *Lying and Its Detection* had attempted to be), as it was a collection of extraordinary claims and sensational anecdotes.[23] About the Lindbergh baby abduction and murder case, for example, Marston hoped to find "a living human being whose mind contains information about the Lindbergh kidnapping. If such a person exists, his secret knowledge can be read like print by the lie detector."[24] Only someone with Marston's arrogant exuberance could have opened a book by praising the Divine's scientific acumen in the Garden of Eden: "God's method was wholly scientific. He observed the suspects' behavior and reasoned logically that this behavior was an outward, visible expression of hidden emotions and ideas of guilt which the man and woman were attempting to conceal. This is the true principle of lie detecting. From that first successful lie detection at the dawn of human history to the discovery of the blood-pressure test for deception in the Harvard Psychological Laboratory, A.D. 1915, millions of human beings have attempted in thousands of different ways to apply this detection principle to specific cases of deception."[25] The "discovery" of the systolic blood pressure deception test in "A.D. 1915" finally ended "the 6000-year search for a truth test."[26]

Given his ambitions for the lie detector, it was no wonder that Marston invested its discovery with such portentous significance. Crime had reached epidemic proportions by the mid-1930s according to him, and it was costing the nation a quarter of its annual national income.[27] Nevertheless, there was a

"It is a psychological medicine," Marston said of his deception test, "which will cure crime itself when properly administered." Frontispiece to William Moulton Marston, *The Lie Detector Test* (New York: Richard R. Smith, 1938).

simple solution to this complex problem: "The deception test, or 'Lie Detector' as it has come to be called, is not to be regarded as one more tool in the police kit for making routine detective procedure a little more effective ... the Lie Detector goes to the heart of the situation. It is a psychological medicine, if you like, which will cure crime itself when properly administered."[28] The root cause of crime was psychological. Because the essence of criminal nature was the power to deceive and the habit of deception, the purpose of lie detector was to "break down all the habits of lying and build up instead mental habits of telling the truth." "The ultimate use of the Lie Detector"—"a kind of psychological X-ray capable of destroying the cancer of crookedness wherever it takes root"—was not for crime detection but for crime elimination: "For criminal investigators the Lie Detector supplies a master key capable of unlocking that vast storehouse of secret information, the human mind, hitherto impregnably protected by an invulnerable wall of deception."

Not only was it an efficient servant for prosecutors, police, and taxpayers alike, but the instrument could also assist with solving marital problems. "Only the *truth* can bring about a real emotional adjustment," Marston wrote. "Deception always destroys love and happiness even though the lie be told from the finest of motives." Problem children could be adjusted by testing their parents with the machine. Such a case was that of one Bobbie K., a troublesome six year old. The child had temper tantrums, was unruly at school, disobeyed his parents, and frequently ran away from home. The boy's mother was undoubtedly over-indulgent, Marston reported. "But Mr. K. made up for that by sterner disciplines including occasional spankings which were the only punishments that seemed to have any effect on Bobby." The psychologist thought that there might be some fundamental emotional conflict in the home, "reflecting itself in the child's behavior as parental disturbances always do." Both parents, however, denied having serious marital quarrels and professed ardent love for each other and for their child. Having persuaded him to submit to a lie detector test, Marston discovered that the father resented the boy as a barrier between himself and his wife in their hitherto passionate relationship. "I forced the father to acknowledge the truth. I did this as brutally as possible and the truth shocked him into adjustment." Fourteen months later, Marston reported, the child's behavior began to improve, the parental re-adjustments considered a success.

Looking to the future, Marston envisaged three possibilities for the lie detector: in politics, in marital and domestic affairs, and in supplying a motive for moral education. "Suppose every candidate for public office had to take a Lie Detector examination on his past record before his name went on the ballot," he suggested. "Suppose every District Attorney had to take a test every six months, as bank officers do where the deception test system is in operation. Suppose governors, mayors, and lesser political office holders had to submit to Lie Detector examinations periodically concerning their use of the tax-payers' money and their own personal contacts with racketeers, known criminals who somehow had always escaped prosecution, and other notorious representatives of predatory interests and the underworld. Suppose the results of these tests were made public automatically, by law."[29] Marston was advocating nothing less than a complete interweaving of the social fabric by the lie detector.

For Marston, the instrument was a means to an end; for Leonarde Keeler it was an end in itself. With no adjunct psychological project to promote, Keeler was devoted to developing scientific lie detection throughout his life. If any one person could be held responsible for furthering its cause in the United States, it was he. Keeler had come from a family that had been very much part of the Berkeley artistic and legal establishment. His father, Charles Augustus Keeler, was an eminent man in his own right, a civic leader, and writer of poetry and popular books. In 1909 he embarked upon a three-year world lecture tour when his son was six years old. At some point prior to the First World War, he arranged for Leonarde to live with the family of the daughter of Frederick Adams, judge of the Supreme Court of New Jersey. An important and well connected man, Keeler Senior was Director of the Berkeley Chamber of Commerce during the 1920s, and by the early 1930s he had turned to writing scripts for radio plays, committed to producing one a week.[30] Following his father's example, the young Leonarde gave radio talks, on one occasion on the subject of rattlesnakes, becoming an "instant celebrity" his sister later recalled.[31]

August Vollmer, chief of the Berkeley Police Department since 1909, was a family friend of the Keelers. When Vollmer was but a mailman, Leonarde would ride around Berkeley in the basket of Vollmer's bicycle. It was through Vollmer that the young Leonarde became involved in criminology. Vollmer regarded Keeler "with the affection that a father would a son," and he was an important influence on Keeler throughout his life.[32] A knowledge of Vollmer's professional project is crucial for understanding Keeler's subsequent career.

August Vollmer has been described as one of the most significant figures in the history of American law enforcement.[33] Influenced by his extensive knowledge of European criminology, Vollmer set about first modernizing the Berkeley department, and later the nation's police.[34] His goal was to create a professional and committed body of law enforcers. Such a program required not only the recruitment and training of highly motivated and intelligent officers, it also needed a supportive bureaucracy of centralized and efficient record keeping. Vollmer established the first formal training for officers and the first crime laboratory, employing intelligence and psychological selection tests. He introduced the modus operandi method of crime analysis, and, in 1925, set up a Crime Prevention Division, recruiting the first social worker to an American police force.

Since the beginning of the twentieth century, rehabilitation had replaced punishment as the ambition of criminological reformers. By advocating a social work function for the police, Vollmer was in line with the latest progressive thinking, as the title of his 1919 article for the *National Police Journal* evidenced: "The Policeman as a Social Worker."[35] Not only would the policeman be an effective crime fighter, but he would also attend to the processes of crime prevention through his personal dealings with the community. To this end, he introduced a number of innovations such as employing policewomen, attending to juvenile crime, and prohibiting the use of violence against suspects—the so-called third-degree. Courses in abnormal and criminal psychology became mandatory for trainee officers, taught by faculty members from the University of California. One officer recruited from the university succinctly summed up the new role when he asserted that "you're almost a father-confessor; you're to listen to people, you're to advise them."[36]

In addition to serving the community, Vollmer argued the police should arm themselves with science. For this reason he has consequently been called "the father of modern police science."[37] In his 1921 presidential address to the International Association of the Chiefs of Police, he proposed that the police should become "armed with facts, not fancies, and with a constructive program for the mental, physical and moral health of the subject."[38] The policeman was to become a "practical criminologist." Inefficient trial-and-error methods of the past must be replaced, he said, with those employed by "microscopists, chemical analysts, medicopsychologists, and handwriting experts."[39] "With very few exceptions," Vollmer wrote, "our archaic system is responsible for deplorable conditions." "Our whole method must be abolished before we can succeed. We must devise scientific methods and apply them to the investigation and removal of social, economic, physical, mental, and moral factors underlying crime and vice. Prevention and not punishment must be our ultimate objective. We must develop experts who have intelligence, training, and character, and they must employ the best scientific and professional tools."[40] Police officers must have special qualities of "intelligence, tact and sympathy," as well as a knowledge of "medicolegal evidence, and even of general medical and psychological principles" and scientific investigation.[41]

Vollmer's drive towards professionalism was a welcome message during the progressive era, concerned as it was with rising crime rates. Crime was a primary public anxiety during the decade of prohibition and well into the 1930s, a concern Vollmer astutely exploited in his dealings with the press. Public support was crucial if his reforms were to attract funding and become law. Crime control was an important political issue during the 1920s and 1930s.[42] Although a competent politician in his own dealings with the press, policy makers, and the wider community, Vollmer maintained that a professional police force must be divorced from politics. Politics was a major threat to professionalization; neutral autonomy was the key concept in his new bureaucracy.[43] Such ideas resonated with a cynical electorate disgruntled with corruption and dismayed at the political influence on police appointments. A professional police force, Vollmer argued, would simply be concerned with the disinterested pursuit of truth.

It is not surprising then that Vollmer encouraged the construction of the first lie detector to be employed in a police force with any regularity. The perfect symbol of the new scientific and professional ethos, it was also a marvelous device with which to attract publicity, a Trojan Horse for his reformist agenda. John Larson, at the time one of Vollmer's "college cops," was detailed to build the instrument, and he was to be helped by Keeler, by then a seventeen-year-old high school student.[44] The instrument embodied many of the values central to the emerging philosophy of professional policing. It was a technology that exploited the wisdom of the social sciences, and although also an interrogation device of sorts (more often than not eliciting confessions prior to the examination), it was evidently ethically superior to the third degree. Attending to the interior life of the individual, and claiming to ascertain truth with scientific accuracy, its objectivity symbolized the new ideals of policing, committed as they were to ending corruption and incompetence. The objective status of the instrument mirrored the police's putatively apolitical status. In this regard it fulfilled a crucial public relations function. It was a potent symbol of the progressive era sensibility, a sensibility that, at its most optimistic, aimed to transform society.[45]

When Vollmer was invited to transform the Los Angeles police as he had Berkeley's, Keeler immediately followed. "It is a pleasure to us to know he is where he comes under your inspiring influence," wrote Leonarde's father to Vollmer, "and we are continuously grateful to you for all you have done for him."[46] Vollmer and his wife would come to regard Leonarde as their "adopted son." Keeler in turn would visit "the Chief" at every opportunity, and would maintain a lifelong correspondence with the man he regarded as his mentor. After his father's death in 1937, Keeler wrote to Vollmer to express thanks for his condolences and support: "I suppose it sounds foolish for a guy my age to say it,—but you know when a father goes one looks to someone else to take his place—and of all the people in the world, Chief, you're it."[47] The young criminologist would later address Vollmer as "Dad Chief."[48] He even named his pet dog "Chief," after "America's Number 1 crime fighter."

Keeler socialized with law enforcers and Hollywood entertainers alike. During the 1920s he met stars such as Charlie Chaplin and Mary Pickford, and he was a long-time friend of Cecil B. DeMille's niece, Agnes de Mille.[49] The law and Hollywood would become the polarities between which Keeler would forge a personal style, an uneasy ambivalence the lie detector would also embody. Although it was a scientific crime-fighting tool it was also sensational and newsworthy. One of the earliest magazine articles about the lie detector expressed this double-identity well. The 1924 *Collier's* piece "The Future Looks Dark for Liars" was illustrated with a photograph of "The Lie Detector in Action." Keeler can be seen operating the primitive-looking machine, which sits on a table. Significantly, the author chose to use a theatrical metaphor to describe the novel scientific instrument. After all, Keeler had—significantly—built some of his early models in a shack on an empty Hollywood lot.[50] And despite claiming it had no "vaudeville features," the journalist was obliged to make some sensational claims about "this strange machine" that was about to make crime almost impossible. "We never found it to err," Keeler said. "A very proud, tired, keen-eyed young man, Leonarde Keeler!"[51] The "youthful inventor" spent the rest of the 1920s working for Vollmer, refining his instrument and technique and attending university. In 1925 (a year after *Collier's* had reassured its readers that "the high-spirited young men who are living night and day with the real lie-detecting idea" were "altruistic real scientists" "with no commercial ambitions"[52]), Keeler attempted to patent a mechanism that allowed him to market the "Keeler Polygraph." It was a prescient move: by the late thirties he would find it necessary to fully commercialize his activities.

Keeler's first real professional opportunity came in 1929 with the founding of the Institute for Juvenile Research in Chicago. Through Vollmer, Keeler was introduced to the Institute's director, Herman Adler, who offered him a full-time job. The position allowed him to perform experiments with the lie detector, and in 1930 he published his first research paper, "A Method for Detecting Deception."[53] Although he recognized that the Institute post was an important opportunity, Charles Keeler attributed his son's poor university performance to his spending too much time with the lie detector. "I am afraid he is a little too indifferent about degrees," he added. Nevertheless, he asked Vollmer if Leonarde's work with Adler could count toward his Stanford degree.[54] A short while after taking up the Institute for Juvenile Research position, however, a new opportunity arose. Appalled by the 1929 St. Valentine's Day massacre, two Chicago businessmen, Burt Massee and Walter Olson, had resolved to counteract what they perceived as a rising wave of criminal activity in the city. With the encouragement of Dean John Henry Wigmore, the nation's principal expert on the law of evidence, Northwestern University's Scientific Criminal Detection Laboratory opened in 1930. Science was to lead the fight against crime. Ballistics expert Colonel Calvin Goddard was appointed director of the laboratory, and Keeler was hired as the resident polygraph operator. Keeler's sister later recalled that while he had a great deal of experience by this point, he was still "waiting in the wings." The "stage was set, the actors ready," she wrote, "but the curtain had not yet risen on the big show."[55] The theatrical metaphor was appropriate. Not only would his successful cases be widely publicized in the newspapers and on the radio, but Keeler himself would soon perform on the national media stage.

Not long after he had taken up the Chicago position, the *New York Times* reported that the lie detector had been used to solve a bizarre but typically newsworthy case.[56] William Tobin, a policeman, had been responsible for looking after a valuable trick canary—the only estate left by a woman who had committed suicide. Tobin shirked his responsibilities, however, because a few days after the bereavement, the dead bird was found in a pile of rubbish. Keeler was able to solve the case by subjecting Tobin to a lie detector test. The suspect, who "displayed marked tension regarding the disappearance of the canary," eventually confessed to killing a cheaper bird and placing it in a dark corner of the apartment. The incident provided Keeler with an opportunity to publish his second paper in the *American Journal of Police Science*, "The Canary Murder Case."[57]

Although most of the events reported by the press were rarely so trivial, the lie detector was invariably associated with the more sensational and newsworthy aspects of crime fighting. In 1931, Keeler's instrument cleared one of Al Capone's lieutenant's, Gus Winkler, of involvement in a $2,500,000 bank robbery.[58] In 1933, Joe Blazenzits was freed from the Marquette Penitentiary in Michigan, having passed a lie detector test conducted by Keeler. Convinced of his innocence, two of Blazenzits' female correspondents had campaigned for the convict's release for five years. Having read about the instrument in a magazine, the women considered the Keeler polygraph their last chance at freedom. According to Eloise Keeler, one of the women later married the former inmate.[59] Keeler was regularly approached by people who thought his instrument could help solve their own problem cases. In 1934 the wife of the convicted Lindbergh baby kidnapper Bruno Hauptmann contacted him to see if a lie test could help in his case.

In addition to giving real lie detector examinations, Keeler also gave demonstrations to organizations such as the Police Lieutenant's Association, Dean Wigmore's Northwestern University law class, the Lakeview Lions' Club, and the Women's Aid Society. In 1933 he received a Distinguished Service Award from the Chicago Junior Association of Commerce for making "the most outstanding civic contribution to Chicago" the previous year.[60] He was credited with obtaining fifty-four direct confessions out of 627 deception tests of crime suspects. "The steadily growing number of cases on which Mr. Keeler is called," the spokesman for the award committee said, "shows that his work is winning increasing recognition." "In fact, it is no exaggeration to say that the laboratory is fast becoming a second Scotland Yard."[61] Eloise Keeler dates Leonarde's attainment of "acclaim and fame" to this incident: "From now on, he would no longer be the boy wonder or the "youthful" Leonarde Keeler. He was recognized now as the authority in his field. He'd have friends galore—many who loved him for his charm, his gift of storytelling, and way with people. Others would want something or would envy his popularity."[62]

The Scientific Crime Detection Laboratory provided Keeler with a base from which he could develop his machine. If the lie detector symbolized the dreams of criminology, and Keeler was the instrument's guardian, then it was ironic that a man with such minimal qualifications should come to assume—in the public eye at least—the mantle of the nation's chief criminologist. But the new criminology was exciting and newsworthy, and the photogenic young man was confident dealing with fellow cops, journalists, and radio producers alike. Keeler's fame increased steadily throughout the 1930s, partly because his students and followers depicted the "Keeler Polygraph" as synonymous with the lie detector. In a 1935 *Scientific Monthly* article, Fred Inbau explained that there was only one reliable kind of instrument, the "Keeler Polygraph." "An instrument of this type should be distinguished from the numerous other so-called 'lie-detectors' frequently found in the psychology departments of many universities," wrote Inbau, one of Keeler's early students.[63]

The ability of the instrument to symbolize the new criminology was also a factor that led to Keeler's involvement in another publicity stunt, the hapless "Illinois State Police Mobile Crime Detection Laboratory and Emergency Unit." Keeler worked on the ill-fated venture with T. P. Sullivan. The two men, longstanding friends who had worked on many cases together, "received nationwide publicity" for their creation.[64] Brought into service in 1942, the "Mobile Crime Lab" was an armor-plated bus crammed with technical equipment. "It is almost a complete crime detection laboratory on wheels," Keeler enthused, "a rolling police department with better equipment than the majority of departments in the country; a mobile hospital where two major operations can be performed simultaneously; a combat unit tough enough to handle anything short of an army tank; an emergency electric power station, radio transmitter, fire fighting and life-saving unit; in general, a complete emergency outfit designed to bring the finest in crime detection, law enforcement, and life-saving facilities to any part of the state on a few hours notice."[65] Some publicity film of the mobile crime laboratory in action survives.[66] The "speedy juggernaut of justice" is seen arriving at the scene of a stakeout. Fired upon by a bandit, the bullet-proof turret is raised and a police officer returns fire from an aperture. Overwhelmed by the unit's superior firepower, the bandit surrenders and after being arrested and brought into the mobile crime lab itself, is given a lie detector test. Fingerprint and chemical analysis would doubtless follow, viewers can assume. As if to symbolize its importance to the project of scientific criminology, the lie detector was situated in the middle of the mobile crime lab, at the epicenter of scientific criminology.[67] "Alas, the mobile unit turned out to be a dud," recalled Eloise Keeler. It was too small "for a crime lab or hospital, too big for a mobile unit." Nevertheless, the bus still performed an important function: "it was used mainly as an exhibit at fairs," a pathetic but appropriate final resting place. After all, that which had spawned it, the Scientific Crime Detection Laboratory, was also, to some extent, a showcase enterprise.

In a 1932 *Review of Reviews* article "Science in the Detection of Crime," William A. Dyche first listed the numerous activities of the Scientific Crime Detection Laboratory before focusing on one particular technology: "An interesting feature of the laboratory has been developed through the efforts of Dr. [sic] Leonarde Keeler, one of the first scientists to see the possibilities of the polygraph, the 'lie detector.' In the operation of this machine is found a combination of physics, biology, and psychology, which can be employed in nearly every situation where it is desirable to obtain knowledge which a suspect may have."[68] "It is believed that no more important invention has ever been made for successfully dealing with crime in the whole course of criminal science," Dyche concluded. The article was illustrated with a photograph of Keeler performing a lie detector test and emblematic diagrams of "truth" and "lie" graphs. *Forum and Century's* piece on the "practical and humane" methods of the new scientific criminology opened with the claim that the Keeler Polygraph was perhaps "the most dramatic and satisfactory of these instruments."[69]

Apparently able to commit the guilty and free the innocent, the lie detector was a wonderful public relations tool for the Scientific Crime Detection Laboratory, perfectly symbolizing the objective and humane ideals of criminology. As Keeler himself put it, the polygraph was a "modern scientific procedure, the antithesis of the old third degree method for determining truthfulness."[70] But its appeal was not necessarily self-evident; its enthusiasts occasionally had to go to absurd lengths to make their point. Business was booming in the laboratory during the mid-1930s, recalled Eloise Keeler: "Every type of case one could dream up was brought in, including one involving the purported mummy of John Wilkes Booth. The mummy was being shown at fairs and carnivals, and, after lie tests, the owners 'conceded' it was a fraud."[71]

The instrument's capacity for attracting sympathetic publicity was a function Goddard was well aware of when he asked Keeler to organize the Scientific Crime Detection Laboratory's exhibition stand for the 1933 Chicago World Fair. Despite a laboratory full of ballistics, forensic chemistry, handwriting analysts, and specialist photography, nothing could capture the public's imagination quite as well as the lie detector.[72] A 1934 *Literary Digest* article about Goddard's laboratory was almost completely devoted to the instrument.[73] The machine was more than a lie detecting device; it was a scientific, humane, and moral technology of truth. And because every crime was apparently entrenched behind a lie, the instrument embodied what *Living Age* had claimed was "the constant dream of jurists and police functionaries."[74] The Chicago laboratory provided the inspiration for the creation of many others. Cincinnati established a scientific crime detection laboratory in 1934, stocked with "ballistic and other scientific identification apparatus, such as the lie-detector and equipment for chemical analysis."[75] In 1931, Charles A. Appel was instructed by the FBI to attend Northwestern's course in scientific crime detection before establishing the Bureau's first scientific crime laboratory.[76]

Publicity brought the Chicago laboratory a heavy case load. "Work is mounting up here faster than ever," Keeler told Vollmer in July 1932. "I have had practically no time to do anything but work on cases, for they have been coming in from all over the state."[77] Furthermore, the volume of work had recently forced a relocation to another building: "We occupy the entire third floor," Keeler continued, "and have a real honest-to-goodness show place, as well as an ideal laboratory set up."[78] But in spite of the enthusiasm, the laboratory soon ran into problems. By 1934 the university had become somewhat dissatisfied with its management. Its budget was to be severely cut, and although the director's salary had already been halved, his services would no longer be required. "Of course there were other reasons for dropping the Colonel" excepting purely financial ones, Keeler told Vollmer, "mainly because of rumors reaching the ears of the faculty members about 'wine, women and song.'"⁷⁹ "Although Colonel Goddard is a fine person and we are all very fond of him personally, he is not very discreet about his overt activities, and sometimes the feminine call muddles his judgment a little in business affairs." It was a rather odd comment coming from a man whose own wife fretted that he possessed "a gift of attraction for men and women alike."⁸⁰ Goddard's solution to his problems was revealing, considering the manner in which he had directed the activities of the laboratory prior to his dismissal: "Colonel Goddard opened a small office of his own next door to the laboratory and spent his summer months managing a bally hoo side show at the fair. He had a crime detection laboratory exhibit but, unfortunately, mixed into it crime-horrors, methods of torture, an electric chair demonstration and part of the time had Dorothy Pollock—"Chicago's most beautiful murderess"—as an exhibit in person."⁸¹ "I'm afraid the show venture somewhat hurt the Colonel's professional standing," Keeler lamented, possibly alluding to the fact that Goddard's gun display was burglarized twice during the fair.⁸²

Keeler's letters to Vollmer during the 1930s regularly reported on the dire straits of the laboratory's finances. Its losses were $12,000 a year by 1934, compared to losses of $40,000 "before Mr. Massee lost in the market."⁸³ By the summer of 1936 its activities were increasing all the time, he recounted, and there were some prospects of obtaining endowments.⁸⁴ The authorities were making a concerted effort to raise sufficient funds to put the Laboratory on a sound, permanent financial basis, he told Vollmer in early 1937.⁸⁵ By then, the laboratory's managers wanted to increase its revenue from $20,000 per annum to $52,000 per annum by forging a better working relationship between the laboratory and the State Bureau.⁸⁶ But it was hopeless. The end came in 1938 when the Dean of the University completely cut off the Laboratory's budget and it was sold to the City of Chicago Police.⁸⁷

Although he was disappointed with Goddard's eye for publicity, Keeler was also at risk of turning himself into a one-man vaudeville act. Anxious to disassociate himself from trivial ventures, he acknowledged that the laboratory was inevitably a "show place" for the new criminology, and the lie detector its central exhibit. After being sued for $750,000 following a disastrous ballistics case, Goddard vowed never to take the stand again or deal with fire-arms identification. "He was going to make money now," Keeler reported, "radio programs, stock promotion in patent medicine,—anything. He was through with being professional and ethical; he had decided just to make money, no matter how."⁸⁸ As the finances of the laboratory dwindled during the Depression-racked thirties, Keeler himself was also forced to commercialize his activities. Despite financial problems, he attempted to maintain his integrity, claiming that his "real interest" was "in the study of human behavior and not in inventing and making money out of some instrument." "Of course," he added coyly, "if some money comes along, it will be welcome, but I think that is incidental to the problem."[89]

The lie detector had always provided the laboratory's only real source of income, excluding gifts and grants. "The Polygraph work is going along as usual, bringing in approximately ten thousand dollars a year," Keeler told Vollmer in May 1938, "and the other work in the laboratory consists mainly of research and occasional cases."[90] Acting on Vollmer's advice, and with the help of Bert Massee, Keeler opened his own lie detecting business, Leonarde Keeler Inc. While it would doubtless be a challenge, he had "many clients awaiting the opening of [his] new office" he reported in August 1937.[91] By the fall of 1938 the business was well under way. "I am handling more cases now than were ever handled in a corresponding period at the old laboratory," he told the Chief in November. "My gross income is approximately $1,000 a month which from all appearances will continue more or less indefinitely. As a matter of fact, I have cases scheduled for about a month in advance, and undoubtedly if I sought more work, I could easily double the gross income."[92] Business was so good by early 1939 that he found it necessary to recruit an assistant.

In addition to running polygraph examinations for the police, banks, insurance companies, and large department stores, Keeler also made money by selling his famous "Keeler Polygraph." He also insisted that future operators take his brief course in the detection of deception. In 1939 the Chief of Police at Toledo, Ohio, reported that the city's officials "feel that our Keeler Polygraph for lie detection is an asset equivalent to an increase in personnel and has paid for itself several times over. In a great many cases, it has quickly broken down the alibis of hardened criminals."[93] In September 1939, a second company, Deception Tests Service Co., of Berkeley, California, followed Keeler into the marketplace. Prices for lie detector polygraphs in the late 1930s ranged from $250 for the Berkeley Psychograph to over $1000 for the Darrow Photopolygraph. Keeler's instrument sold for $450.[94]

Although he could no longer operate from within the safe confines of the Scientific Crime Detection Laboratory, Keeler's fame brought him all the work he needed. In February 1941, *Reader's Digest* reprinted a *Forbes* magazine article entitled "The Lie Detector Goes into Business."[95] "For 10 years," the piece began, "many of Chicago's banks, department stores, chain stores and restaurants have been using the Keeler Polygraph, or 'lie detector,' with astonishing results." By the early 1940s, according to the magazine, ninety-five percent of Keeler's work was commercial. In 1944, after the *Saturday Evening Post* published a series of three articles about him and his "Magic Lie Detector," he told Vollmer that the articles were "bringing in all kinds of new business."[96] In 1946 Keeler was instrumental in tracking down $1.5 million worth of missing jewels. "There was enough to fill the windows of half a dozen shops like those on Madison Avenue, New York," reported the Sunday *New York Times*. "As important as the loot, recovered with the aid of a twentieth-century lie detector, were the confessions that officials said had been obtained from principals in the fantastic crime."[97] Two senior Army officers had stumbled upon the Hesse-Darmstadt jewels in the Kronberg castle toward the end of the war and had secreted them to the United States.

Keeler was sufficiently famous by the 1940s that he could play himself in the Hollywood movie *Call Northside 777*. Starring Jimmy Stewart, the film told the true story of a reporter who employs the tools of scientific criminology to free an innocent man from jail. A fine example of film noir, the movie was released on February 18, 1948, some eighteen months before Keeler's death. In many ways it was the peak achievement of a widely celebrated career. By shooting the movie in the original Chicago locations and by adding cinema verité touches like Keeler and his polygraph, the filmmakers were aiming at documentary style realism. Keeler's appearance in the film was an appropriate finale to a career that having started in Hollywood, was destined to end there.

The lie detector was constructed by and in American popular culture. William Moulton Marston and Leonarde Keeler were the two most influential individuals responsible for establishing its use in the United States prior to the Second World War. A tireless popularizer of psychology who created "an entire oeuvre of 'lowbrow' literature,"[98] William Moulton Marston designed *Wonder Woman* to be an embodiment of his esoteric social philosophy. Appropriately described by one historian as having a "mania for publicity," Leonarde Keeler was apparently the inspiration behind that other crime-fighting comic book hero *Dick Tracy*.[99] Both men had a flair for theatricality and were adept at dealing with the press. Both courted opportunities to appear in the media. And both possessed that quasi-magical quality of leadership that Max Weber called charismatic authority.

During periods of relative social and political calm, Weber argued, the habitual demands of ordinary life were brought about by mundane power structures embedded in bureaucracies. Bureaucrats and patriarchs are afforded leadership roles because they embody those rational rules that made everyday governance possible. At times of rapid social change, however, a form of authority based on personal charisma can emerge. During such times, ordinary managers are usurped by leaders who come to inspire intense loyalty.[100] "It is the *duty* of those to whom he addresses his mission," Weber argued, "to recognize him as their charismatically qualified leader."[101] Neither patriarchal or permanent, charisma is a fickle resource. The charismatic hero's power does not arise from formal codes or statutes, traditional customs, or "feudal vows of faith," as in patriarchy.[102] It is gained solely by proving its strength, performing miracles or heroic deeds, and by embarking on extraordinary ventures. The "god-like strength" of the hero makes a sovereign break with all traditional or rational norms. In the case of the lie detector, charismatic authority was intimately tied to the myth of invention that was in turn the source of the machine's mystique and power. Invention was a highly valued commodity within the moral economy of lie detecting, and it played a crucial role in establishing lie detection as a credible and meaningful activity. The title "inventor of the lie detector" conferred status upon its holder and justified and consolidated his charismatic authority. And in turn, the charisma of the invention was itself a crucial component of the lie detector's spectacular powers.

CONCLUSION

The Hazards of the Will to Truth

> The will to truth, which is still going to tempt us to many a hazardous enterprise; that celebrated veracity of which all philosophers have hitherto spoken with reverence: what questions this will to truth has already set before us! What strange, wicked, questionable questions! It is already a long story—yet does it not seem as if it has only just begun?
> —Friedrich Nietzsche (1886)

> "This is not my idea," he said.
> "Yes, Inspector Bryant told us that. But you're officially the San Francisco Police Department, and it doesn't believe our unit is to the public benefit." She eyed him from beneath long black lashes, probably artificial.
> Rick said, "A humanoid robot is like any other machine; it can fluctuate between being a benefit and a hazard very rapidly. As a benefit it's not our problem."
> "But as a hazard," Rachael Rosen said, "then you come in."
> —Philip K. Dick (1968)

"CHICAGO, Tuesday, March 2.—Joseph Rappaport, murderer of Max Dent, a government informer, died in the electric chair early today after a struggle for a reprieve which had its climax when a lie detector was carried into the death cell and scientists, lawyers, jailers and witnesses stared at tiny needles tracing on a slowly moving paper ribbon what proved to be mute lines of doom." Thus begins a 1937 *New York Times* article, "Lie Detector Seals Doom of Murderer."[1] Having won a stay of execution once by order of the State Supreme Court and four times on the order of Governor Horner, Rappaport's hopes lay with the lie detector, a machine in which the Governor had declared himself to be a "great believer." "Judge in the test," enthused the *Times*, "was Professor Leonard Keeler of the Scientific Crime Detection Laboratory of Northwestern University, inventor of the detector, scientifically known as a polygraph."

Keeler and his party had filed into the cell, according to *Newsweek,* interrupting a card game between the condemned man and his guards.[2] The criminologist "put an odd-looking black box-like machine, about two feet square, on the table" and then tightened a rubber tube around Rappaport's chest.[3] The suspect looked tense "as a rubber sack attached to quarter-inch rubber tubing filled with mercury was wrapped around his upper arm and inflated to midway between systolic and diastolic blood-pressures." The interrogation began:

> "Is your name Rappaport?"
> "Yes."
> The stylus flowed evenly across the slow-moving ribbon.
> "Did you kill Dent?"
> "No."
> The pen dashed off an inch-high peak.
> "Is your home in Cook County?"
> "Yes."
> The needles graphed an even course.
> "Do you know who shot Max Dent?"
> "No."
> The needles quivered, flickering jagged lines on the tape—mute lines that sealed a murderer's fate.[4]

"Rappaport went into the death chamber at 12.04," the *Times* reported. "He was pronounced dead at 12.12."[5]

Three months later, on June 8th, 1937, the *Times* told the story of another troublesome case whose resolution rested with the lie detector. Around 5:30 p.m. the previous evening two boys, aged nine and twelve years old, had stepped into a taxi at Lexington Avenue and Forty-second Street in New York.[6] Flashing a ten dollar bill, one of the boys had instructed the driver to take them to a G-Man movie on Broadway. The suspicious driver had instead called a nearby policeman over to his cab. Patrolman Patrick Casey took the boys to the East Fifty-first Street police station for further questioning. While two detectives looked on "with expressions of exaggerated grimness," the patrolman rolled up the boy's sleeve, wrapped a towel around his arm and connected the towel with a string to an alarm clock on the windowsill. The detectives pretended to watch the clock intently, while Casey pointed his finger at the suspect and asked him, "Where did you get that money?" The boys quickly confessed to having found the money at home. They were held at the station until their parents came for them. Both agreed "that their experience with the police had been more thrilling than any G-man movie."[7]

So we have two lie detector tests, one apparently legitimate, the other not. The Chicago test had been conducted on a convicted murderer, with the State Governor's blessing, by the great Professor Keeler, a man widely known as the inventor of the lie detector. The New York test had been little more than a crude police station joke, a shameful charade whose intention was to cajole two bewildered boys into admitting their guilt of petty theft.[8] In Chicago, justice was seen to be done, while in New York proper procedures were abandoned. The Chicago police used a real lie detector, but their New York counterparts rigged up a fake one.

The newspapers may have crowned Leonarde Keeler the inventor of the detector, but he always contended that "there is no such thing as a 'liedetector.'"[9] The polygraph, he maintained, simply recorded the body's physiological changes onto a chart, which the skilful examiner must then scrutinize in order to arrive at a "diagnosis" of guilt. Because a person undergoing a lie detector examination "responds almost continuously to his immediate environment, to other individuals, to sounds, odors, pain, and other stimuli factors," Keeler argued, all attending circumstances must be devoid of "irrelevant factors" if correct procedures were to be followed.[10] Drawing an analogy with the medical examination, Keeler believed that the subject of a deception test must refrain from eating or drinking for several hours, rest quietly for fifteen or twenty minutes before the test, and be undisturbed during it: "examinations must be conducted in darkened rooms, or in quiet environs."[11] Thus, even according to Keeler's criteria, Rappaport's last minute death chamber lie test was deeply compromised by the attendance of a rowdy group of "scientists, lawyers, jailers and witnesses," not to mention journalists: the apparently genuine Chicago test displayed some troubling and dubious features.

Certain aspects of the bogus New York test, however, conformed to orthodox polygraph practice. In his 1930 article, "A Method for Detecting Deception," Keeler argued that suspects should be made wary of the machine and be reminded of the lie detector's capabilities with the following preamble: "This machine to which you are connected has been used for some years on criminal suspects, and so far has proved a very reliable means of detecting the innocence or guilt of a man, and I'm sure we will not fail in your case." A single test of ten to fifteen minutes was usually sufficient to ascertain guilt,

Keeler suggested, but subsequent procedures could be introduced if necessary. A suspect could be shown the machine's polygraphic tracings, for example, and asked to account for his emotional stress. Or a second test could be arranged so that the subject would be forced to watch "the excursions of the fluctuating needle." It was important to utilize a suspect's fear because "the tell-tale needles will only tend to magnify each excursion as he sees them recorded," Keeler explained. By this point about seventy-five percent of the guilty suspects confessed. But if a confession was still elusive, "a night incommunicado" would doubtless assist. "The only 'torture' involved in such a test," Keeler clarified, "is self-induced through fear of being caught, and that fear exists whether the man is being cross examined in the usual way or on a blood-pressure apparatus."[12]

By enhancing the dramatic atmosphere surrounding their lie detector test, by drawing their suspects' attention to the machine, and in attempting to produce a confession, the police of the East Fifty-first Street station were following established testing methodology.[13] The two aspiring New York petty thieves might have been attached to nothing but a towel and an alarm clock, but they experienced many of the melodramatic features of an orthodox lie detector test. Although these two stories appear to be polar opposites, that opposition is quite unstable. The experiences of Joseph Rappaport and John McLaughlin delineate one of the lie detector's most pertinent polarities. The instrument symbolizes the enlightened business of crime detection. However, it is oftentimes little more than a source of the darkly comic. The delectable irony of lie detector discourse is that the legitimate and the illegitimate regularly intermingle: this modern science of truth depends on an old-fashioned art of deception.[14] This is a discourse in which scientists become celebrities, scientific instruments acquire magical agency, and standardized practices allow the nonviolent to become violent. It is a discourse that blurs distinctions between the eye and the ear, the scientific and the spectacular, the endogenous and the exogenous, and the normal and the pathological. The polygraph's power is its ability to maintain credibility while tolerating these essential tensions.

Criminology's first object of knowledge, *Homo criminalis,* was construed as a degenerate and pathological species, a reviled "other." Although the belief that the criminal was a biologically flawed type of person was widely held, the theory was emphatically articulated, studied, and promoted by Cesare Lombroso. His notion of the born criminal dominated discussions about criminality from around 1875 until after the turn of the twentieth century.

The establishment of criminology as a discipline went hand-in-hand with the conceptualization of criminality as an essentially biological problem. The theory was the touchstone that all investigators had to acknowledge, even if only to criticize. Later depicted by Lombroso as having come to him fully formed, in a flash of insight, a wide range of nineteenth-century intellectual and practical projects, including statistics, psychiatry, prison reform, philanthropy, evolutionism, and degeneration theory had, in fact, made his work possible. Lombroso also mobilized classical mythology, history, literature, and the natural world, not to mention anecdote and folklore in support of his thesis. It was out of this effervescent ether that the solid and apparently steadfast figure of criminal man materialized. Criminal man, then, was the solution to criminology's first dilemma: crime would be confronted by invoking a eugenic mode of governance for the pathological body of the criminal.

Criminal anthropology had an eye for an arresting visual image and an ear for a compelling story. Yet behind the loquacious rhetoric lay an insecurity, a tension between the desire to govern and the desire for knowledge, and also between science and common sense. Criminology's charismatic bricolage was unstable, and criminal man's aetiology, prognosis, and even his existence were under perpetual threat. The dilemma between scientific expertise and charismatic authority—resolved by the mythopoetic figure of Lombroso—was one the lie detector would later inherit.

Criminal anthropologists were fascinated by the female body. In accordance with long-standing beliefs about gender differences, women were considered inherently secretive, deceptive, and duplicitous: they could not be trusted. They were believed to be essentially corporeal beings, enslaved by their bodies. An entrenched undercurrent of Western thought posited a binary opposition between the male rational mind at one pole and the female irrational body at the other. Like the primitive savage—also regarded as less evolved than the European male—women were thought to be insensitive to pain and generally less "sensible" than men, possessing ineffectual emotionalities.[15] Whereas women were thought to be in need of governance due to their disruptive biologies, men were thought to be sufficiently rational to be able to suppress their emotional and sexual drives. Criminal anthropology had first sought evidence for criminality in the visible: in tattoos or facial physiognomies or via the assessment of the shapes and sizes of anomalous skulls. But this evidence proved elusive. Having decided that the stigmata of crime might not be written exclusively on the surface of the criminal body, criminology had to look ever deeper inside it for those hypothesized "internal lesions" of criminality. That women apparently exhibited no visible signs of criminality was, paradoxically, another reason why criminology came to regard them as perfect suspects. Criminal anthropology was obsessed with prostitutes; when it came to women, the discipline considered the distinction between the normal and the pathological a matter of quantity not quality. The prostitute's criminal trade depended on invisibility—hidden deeds, concealed emotions—and yet the nature of her business was relentlessly corporeal. As the epitome of femininity, the prostitute also embodied criminality: degenerate and duplicitous yet devoid of external stigmata.

Criminal anthropology postulated that because criminals were savages living in the midst of modern civilization, they were indifferent to the suffering of others, devoid of empathy, and were less sensitive to pain compared to the law abiding. Incapable of displaying normal emotions, the criminal's sentiments were vulgar, crude, and unrefined. Criminology, therefore, constructed emotion as either an absence, a deficiency of the criminal mind, or as a pathology, a deviation from the normal. Once the search for visible stigmata of criminality had floundered, an extensive range of physiological instruments was employed to measure the hidden world of pathological emotions. Invisibility, corporeality, femininity, emotionality: this was the discourse of female criminality, a discourse that overlapped with that of hysteria, the most widely-diagnosed female malady at the time. Emotion was the key because it was both concealed within the female body yet also denigrated as rationality's criminal "other." The female body came to occupy a numinous position within criminological discourse via the prostitute's: "Find a solution to the enigma of women, and you will solve the puzzle of criminality!" By offering a solution to this puzzle, the laboratory study of the emotions inadvertently made possible the later development of the lie detector.

Concurrent with the emergence of criminal anthropology was the growing suspicion that crime was a normal if regrettable feature of society. Criminality was regarded by fin-de-siècle novelists as something distressingly ordinary. Criminal anthropology's naturalization of crime had compromised the age-old distinction between the moral and the immoral. The concept of criminal man as a despised other, a species apart, was quickly challenged by novelists such as Arthur Conan Doyle, Bram Stoker, H. G. Wells, and Robert Louis Stevenson. The effect of these "Gothic" critiques was to undermine the notion of criminal otherness. In literature the human and the abhuman were inextricably mingled. By 1900, American criminology was also beginning to question the concept of criminal man as a separate kind of human being. University of Chicago sociologist Frances Kellor was convinced of the illegitimacy of conceiving of the criminal as a separate type of person. Using pneumographs and kymographs—"tests of respiration" as she called them—Kellor investigated the emotions of female offenders. Unlike her predecessors, who had also used such instruments, she considered her subjects to be normal, not deviant.

The American press was fascinated with the technology of criminology. From 1900 to 1920, a series of "soul machines," "truth-compelling machines," and "machines to cure liars" were described with great enthusiasm. But none of these devices were lie detectors in the sense with which the term was understood after the 1920s, because the predominant ambition was still to understand the pathologies of the intellect, emotional complexes, and, of course, criminal minds. These instruments were focused on studying the aberrant, the abject, the abhuman; their target remained the pathological human—a target that had already been deconstructed by novelists and criticized by sociologists. In the context of eugenics, the dream of criminology was, nevertheless, glimpsed as a real possibility: "There will be no jury, no horde of detectives and witnesses, no charges and countercharges, and no attorney for the defense. These impedimenta of our courts will be unnecessary."

Frances Kellor's association of emotion, criminality, and normality within the context of laboratory instrumentation was a significant development. It was only when criminology came to refute the theory of the born criminal that the idea of the lie detector became a real possibility. This refutation occurred first in fin-de-siècle novels and detective fiction during the first decade of the twentieth century, quite a few years before Marston, Larson, and Keeler began their work. During this period, writers such as Melvin Severy, Cleveland Moffett, Charles Walk, Arthur B. Reeve, Edwin Balmer, and William MacHarg all imagined lie detectors well before criminology embarked on solving the practical problems that turned these fantasies into reality. The writers of scientific detective stories no longer had a role for the born criminal, because their "whodunit" plots featured a range of possible suspects, all of whom were apparently equally culpable. Once criminology had accepted that "every crime was entrenched behind a lie"—and no longer the inevitable byproduct of the diseased "criminal mind"—it followed the novelists into accepting that anyone, not just born criminals, were capable of committing crimes. The lie detector was, therefore, made cognizable by the rejection of the obstacle of the born criminal.[16] Because that obstacle had first been cleared away in literature, it is appropriate to credit the literary imagination with the creation of the lie detector.

Lie detector discourse has always been conflicted. What characterizes it is not, for example, an uncompromising attempt to amass empirical evidence to refute its reputation for brutality. Although use of this new technology promised to replace the third degree with scientific humanitarianism, an element of intimidation from the sweat box had its uses. Liberation and subjugation coexist with this technique. Although the instrument was often depicted as an automatic truth machine, the attending human expert could never become entirely obscured. The "charismatic authority" of the expert—to invoke a deliberate oxymoron—was vitally important to the success of the enterprise. The discourse is best understood as one that posits an interrelated series of dilemmatic oppositions between the broadly scientific and empirical and the largely theatrical and performative.[17] The lie detector emerged when and where it did because the cultural conditions were right. The technology could not have emerged in Britain because "criminal man" couldn't establish a foothold there and also because even toward the end of the nineteenth century, lying was not regarded as particularly problematic by the professional and bourgeois classes.[18] In America, however, a Puritanic intolerance of lying (or, at least, a publically proclaimed intolerance of lying), together with a progressive faith in the utopian potential of technology, coincided with the professionalization of the police and the growth of a crime-obsessed sensationalist media. These forces, together with the democratic privileging of the lie as the central problem for criminology and detective crime fiction alike, created conditions for the emergence of the lie detector. If the technology was "socially constructed", then it was so thanks to those excursions that were made back and forth across the boundary demarcating criminology from the wider culture. The instrument's architecture, in other words, lay on the fertile boundary where criminology interacted with its public. In this sense, the lie detector was inherently transgressive. It is not so much that the lie detector was created as a result of "boundary work"—the attempt to demarcate science from nonscience[19] —but rather that a series of deviations back and forth across boundaries—between nations, continents, disciplines, genres, genders, modalities of power, and so on—created the discursive space out of which the practice finally emerged. It was the persistent crossing of boundaries that was so productive, rather than the creation of those boundaries.

Modern criminology was itself created at the intersection of two enterprises: a "Lombrosian project" and a "governmental project."[20] The Lombrosian project proposed that the solution to crime lay in correctly demarcating criminals from noncriminals. The governmental project proposed that in order to administer justice in an equitable way, power must be deployed to take into account the nature of the criminal. Criminology thus emerged out of a double predicament: the solution to the dilemma of governance could only appear once the dilemma of the nature of the criminal had been confronted. The dilemmas of science and governance, then, represent criminology's two foundations. The history of criminology, as well as that of the lie detector, can be understood in terms of how subsequent developments within the discipline posited different solutions to these tensions.

Lie detector discourse was characterized by several interrelated dilemmas. One set, focused on the nature of science, asked, "What is the appropriate object of knowledge?" and "How should knowledge be created?" The other set, focused on the nature of power, asked, "What is the appropriate target of power?" and "How should power be deployed?" Different actors responded to these questions in accordance with their own circumstances, sympathies, and positioning within the discourse. Those who were involved with lie detection had to confront these contradictory impulses.

Constructed on the borderlands between criminology and the wider culture, the lie detector retained its Januslike ability to look in two directions at once. Amid the smoke and mirrors, one thing is certain: the lie detector was not the invention of any one individual. It is not credible to think of it as a scientific "invention" at all. Most of the machine's constituent parts (the sphygmomanometer, the galvanometer, the pneumograph, and so on) had long been in use within criminology, where they had been used to investigate the multiple pathologies of criminal man. Nevertheless, the myth of "invention" played a fundamental supporting role within lie detector discourse. The term that provided invention's dilemmatic opposite—tradition—was only deployed by the machine's advocates when discussing the history of the lie. Whereas the lie was an ancient vice, they argued, the lie detector was modern. To some extent the lie detector was an artifactual creation of the term "lie detector."

All inventions, real or not, require an inventor. Two exuberant personalities came to embody this aspect of the technology: the criminologist Leonarde Keeler and the psychologist William Moulton Marston. These charismatic individuals skillfully resolved some of the lie detector's dilemmas of science and governance. Like all the instrument's advocates, they emphasized

Table 1. The Lie Detector's Discursive Architecture

Dilemmas of Science		Dilemmas of Governance	
What is the object of knowledge?		*What is the object of power?*	
the normal	the pathological	human	inhuman
the lie	the liar	the mind	the body
detection	confession	male	female
internal	external	rational	emotional
social	natural	machine	agency
How should knowledge be created?		*How should power be deployed?*	
truth	profit	invention	tradition
expertise	charisma	discipline	spectacle
laboratory	theater	liberation	subjugation
methodology	magic	trust	suspicion
language	vision	therapy	tyranny

Note: The response of different historical actors to these dilemmas was a function of where they were positioned in the discourse and what their prior commitments were.

the instrument's scientific credentials and promoted its use as an alternative to the brutalities of the third degree. They privileged the chart as the locus of truth and described the machine in terms of metaphors derived from the law courts. Both thought the instrument could assist in the administration of business and politics. But Keeler and Marston, and indeed Larson too, confronted these dilemmas in different ways. They took different positions on whether the instrument was a magic black box (as Keeler emphasized), whether it could intervene in romantic affairs of the heart (as Marston claimed), or whether it had a diagnostic role to play in medicine (as Larson suggested).

In his insightful paper on the instrument's role in solving the problem of trust in twentieth-century America, Ken Alder argues that the history of the lie detector "is part of the history of how America coped with the rise of a mass public, on the one hand, and the rise of new large-scale organizations on the other."[21] Alder suggests that the technology was part of a wider discourse that was itself an "uneasy hybridization" of two different strategies of expertise. The first strategy, open science, valued the disinterested pursuit of knowledge and its dissemination via meritocratic academic channels into the public domain. The second strategy, proprietary knowledge-making, took social utility as its starting point but sought to sell high value knowledge, expertise, or power to interested authorities. While Keeler embraced the latter strategy by patenting and marketing the "Keeler Polygraph" and a training course to go with it, Larson adopted the former strategy by using the instrument as a tool of psychiatric diagnosis. When, in 1924, for example, Keeler had informed Larson that he thought the lie detector was "altogether too much in its infancy to start anything in a commercial way," Larson had responded by praising Keeler's restraint and noting that science and profit were in opposition: "You did right to keep out of the commercial proposition," he wrote, "for I think it would ruin you scientifically."[22] Although Larson would later dismiss the commercialization of polygraphy as "unethical," the two approaches were not mutually exclusive in that "each strategy depended on the other, and each was wracked by internal tensions not easily overcome."[23] Keeler could complain that Marston's work was scientifically worthless, as Fred Inbau, a close associate of Keeler, could claim that the Harvard-trained psychologist Marston lacked scientific credentials. In response, Marston could casually dismiss the myth of invention while claiming to have invented a technique of his own. As he was marketing proprietary knowledge in his populist book, *The Lie Detector Test*, he was claiming credibility for having discovered a scientific principle.

Whether the epistemological impetus was to produce truth or to generate profit, here was a dilemma that different actors responded to differently, depending on their prior commitments. Early in his career, for example, Marston had pursued an open science strategy as he attempted to establish an academic career for himself based on his experimental research on the emotions. From the 1930s on, however, as he moved more into the domain of popular psychology, he adopted the proprietary knowledge-making strategy, positioning himself as a "consulting psychologist," a priestlike purveyor of esoteric psychological knowledge. In claiming to have discovered an important technique of lie detection—and not laying claim to have invented the instrument—Marston was attempting to underpin his populism with scientific respectability. Wonder Woman, his greatest creation, embodied this dilemma of expertise. Although she professed to have no special supernatural powers (her athletic abilities being the result of sheer hard work and dedication), she was adept at operating mysterious technologies that law enforcement authorities were keen to exploit. Her Golden Lasso of Truth was a form of esoteric proprietary knowledge, yet it produced truth in a pure and systematic (albeit mysterious) way.

Marston's advocacy of the lie detector-as-therapy was unique. He maintained that the technique could become a tool of psychotherapy for families in crisis and insecure lovers. Having detected the subconscious secrets of the subjects, Marston would then confront them with the results. Forcing them to acknowledge their "repressed feelings" would be therapeutic: the truth would set them free. Neither Marston nor any of the machine's advocates ever realized this ambition. Its therapeutic liberating potential was restricted to popular psychology texts and speculative magazine articles. The lie detector did not have the psychotherapeutic potential for governing the self, because that project required the willful and enthusiastic consent of its subjects. *Collier's* assertion in 1924 that there was "no immediate danger of the lie detector following the talking machine and the radio set into the intimacy of domestic life" remained true for the whole of the instrument's history.[24]

When he drew up the plans for *Wonder Woman* in early 1941, Marston introduced many of the psychological ideas he had developed throughout his career into the comic's moral economy. In addition to structuring her cosmos between the polarities of dominance and submission, he also equipped his heroine with a lie detector of her very own, one that encapsulated his utopian philosophy of psychology. Should any of her enemies become captured by the Golden Lasso of Truth, they would find themselves incapable of lying. Fashioned from "fine chain links" from Queen Hippolyte's magic girdle (itself a constraining garment), the lasso was "as flexible as rope, but strong enough to hold Hercules!"[25] In the 1944 adventure, "The Icebound Maidens," for example, Wonder Woman used the golden lasso to compel the scheming Prince Pagli to explain his devious motivations, thereby allowing Wonder Woman to free his captives. Like the equally mythic lie detector upon which it was modeled, the lasso was intended to be one of Wonder Woman's principal weapons against the forces of crime and injustice. Wonder Woman would instantly lose her special powers were she to become trapped in her own lasso. For Marston, liberation and subjugation were an essential tension, different sides of the same coin.

As the golden lasso evidenced, Marston was aware of the lie detector's dual qualities as an instrument of liberation and domination.[26] He, therefore, acknowledged a feature of the lie detector that very few advocates were prepared to admit: despite its reputation for scientific humanitarianism, it was a coercive and illiberal technique. Marston believed that the price for obtaining freedom from truth was submission. Although Wonder Woman's community was set on "Paradise Island," he also provided the Amazons with "Reform Island," a penal facility where women prisoners learned "ways of love and discipline"—two categories that, for Marston, were not in opposition. On Reform Island, Wonder Woman's sisters transformed "through discipline and love, the bad character traits of women prisoners."[27] Every prisoner on the island was forced to wear a magic Venus girdle, a belt designed to make the wearer enjoy living by peaceful principles and to "submit to loving authority." Marston recognized that the lie detector was the center of an ideological dilemma that had freedom at one pole and subjugation at the other.

It was clearly a tool promoted and possessed by those in authority: the police, the state, private businesses, and so on. The lie detector test, one might argue, was another disciplinary technique in the arsenal of "technologies of the self" held by those authorities whose responsibilities include classification, regulation, and normalization.[28] But this only captures part of the story. Although this repressive interpretation certainly delineates some useful orienting lines of perspective, it misses some notable features of the machine's modus operandi. Use of the instrument disciplines those unfortunate enough to be subjected to it, but that is not all it can do. Marston suggested that the instrument could be used as a "love detector," a therapeutic tool in relationship counseling, and Larson never abandoned his belief that the instrument could be used in psychiatric diagnosis. Rather than being an exclusively coercive technique, the machine had the potential to cure, to heal, and to encourage. And it could nurture freedom. What could be more liberating than a technology of truth, especially one that promised to reveal affections of which subjects themselves were unaware?

Michel Foucault began his paradigm shifting *Discipline and Punish* by contrasting two forms of power: capricious, violent sovereign power and institutionalized, anonymous disciplinary power.[29] Having described the bloody spectacle of a typical mid-eighteenth-century display of torture and execution, he then presented a series of meticulous codes that were regulating the actions of young prisoners eighty years later. Whereas the former regime used tyrannical sovereign power, by the time the latter were being used the social contract was in place and, in France, a new egalitarian relationship between the state and its citizens had been forged. Foucault traced the shift from a jurisprudence centered on the charismatic authority of the king to one in which numerous controlling mechanisms had been distributed anonymously throughout society. Two emblematic forms symbolized the shift from one regime to the other: the dark dungeon in which prisoners were left to rot at the king's behest and Bentham's "Panopticon" prison design, which aimed at their enlightened rehabilitation.

But Foucault's dichotomy between the spectacle of public punishment and the disciplinary prison, it has been argued, overlooks the similarities between the two modalities.³⁰ Because spectacular punishment and disciplinary panopticism are both mediated by the imagination, both require the distribution of semiotic codes to function. Jeremy Bentham incorporated theatricality into his prison designs: "lose no occasion of speaking to the eye," he wrote. "In a well-composed committee of penal law, I know not a more essential personage than the manager of a theatre."³¹ Prisoners should experience "a permanent subjection to the conditions of being onstage, albeit with none of the sense of an approving audience."³² Prisoners should be led by their reason to imagine their own surveillance within the panoptic prison. Because the Panopticon produced its effects through fictional means, its success was not founded on its materialization: it didn't have to be built to be effective.³³

Theatricality is not an unusual element in the discourses agitating for reform of punishment, even those ostensibly effecting transformations from spectacle to discipline. It was the perceived ineptness of sovereign power's myth making that directed calls for the reform of punishment. Already part of an increasingly public and theatrical court process, English punishment did not replace, but instead transformed those spectacular strategies applied to punishment.³⁴ Thus against Foucault's stark (but rhetorically charismatic) demarcation between the spectacular violence of sovereignty and the routinized regime of discipline can be counterposed an account of the massive production of a highly public image of the law through rich scientific and literary narratives of criminality.³⁵ *Homo criminalis* was nothing if not charismatic. Penal and criminological thinking has always contained spectacular elements. The authority of the modern bureaucratic state materialized in the disciplinary settings of bureaus of records, circulars such as the *Police Gazette*, newspapers, court reports, in the reign of rules and regulations, and in the designs for prisons has been fully humanized only through illusionism.³⁶

In his analysis of the guillotine in postrevolutionary France in the 1790s, Philip Smith finds a continuing role for symbolism in popular, political, and expert discourses on punishment.³⁷ Although the elevated angled blade was intended to provide a scientific, humane, and egalitarian form of execution—reflecting the Enlightenment's cult of reason, efficiency, and novelty—the instrument was also a deeply mythical and totemistic object, a ritualized and magical device. The guillotine's advocates failed to create an authoritative self-contained punitive technology devoid of ambiguous significations. Once released into the public domain, the guillotine's definitive meaning became contested within a discourse of images and symbols.³⁸

The symbolic and mythic qualities of punishment and disciplinary technologies have been overlooked. The lie detector's primordial symbols and mythologies did not arise later, post hoc, but were essential parts of the discourse from the beginning. Representations can have a constructive power as well as merely reflect the order of things after the fact; metaphors can fabricate reality while they translate. Essentially a semiotic technology, the lie detector was a network of signs demanding interpretation, a "book to be read." This "Golden Lasso of Truth" signified many things. It represented the authority of the superhero whose powers were magical. It was threatening and coercive. It promised to eradicate crime. It encapsulated the notion that the price of freedom was slavery.

One function of the spectacle is to conceal contradictions.[39] Lie detector discourse was inherently dilemmatic. Although it was an apparently humane technology—insofar as it was designed to replace the third degree—it also threatened violence. Although ostensibly gender neutral, a strict gender demarcation undercut its workings: the male gaze scrutinized the female body. Although the discourse appealed to science for legitimacy (through its instrument fetishism, the accuracy statistics and graphs, and the pictures of "inventors" wearing white coats), the practice required theatricality to function. Although the discourse privileged the abilities of the instrument to detect hidden lies using scientific instruments alone, suspects' external behaviors and demeanor had to be scrutinized before a diagnosis of guilt could be obtained. Although the machine was depicted as an impassive, automatically-functioning scientific instrument, it could acquire magical agency whenever necessary. The detection of discrete emotion was often presented as the sine qua non of polygraphy, but the most desirable outcome was inevitably a verbose confession.

As in a sovereign technology, the ambition of the lie detector's advocates was the securing of an admission of guilt. But like a disciplinary technology, it rendered subjectivity calculable and promised scientific objectivity somewhat at a distance from the authority of the police. The Golden Lasso's foundational axiomatic paradox was that truth will bring freedom, but truth must be obtained coercively. The essence of the lie detector is neither its promise to produce freedom nor its threat to oppress. Rather the integrity of the lie detector is captured by the dilemmatic choice between the liberal and the illiberal. Considering all these structural antagonisms, it is appropriate that the logo of the American Polygraph Association ("Dedicated to Truth") is essentially dilemmatic: because Justice wears a blindfold she is incapable of interpreting the polygraphic scroll she holds in her hand.

The twin dilemmas the lie detector inherited from criminology concerned how to do science and what to govern. It is not that criminology is a spectacular science; rather it is that criminology's dilemmas of science and governance lead to two contradictory impulses, one undermining the other. Science aims at truth but governance requires spectacle. Criminology becomes trapped in an antagonistic circuit between the will to truth and the will to power. The greater the promise of the new technologies, the more they capture the popular imagination. The more the public clamors for solutions to the problem of crime, the greater the pressure that comes to bear on criminology. The constant antagonism between the scientific and the spectacular is the principal dilemma to which criminology has been subjected throughout its short history. The most successful figures in the history of criminology have therefore been those individuals whose charismatic authority has enabled them to negotiate the boundary between the scientific and the spectacular.

This chapter opened with a quotation from Philip K. Dick's science fiction novel *Do Androids Dream of Electric Sheep*.[40] Early in the story, which is set in 2019, the Blade Runner Rick Deckard is called upon to locate and "retire" a number of Nexus-6 "replicants" who have recently escaped from an off-world colony. Echoing Nietzsche, the bounty hunter Deckard tells the android Rachael, "A humanoid robot is like any other machine; it can fluctuate between being a benefit and a hazard very rapidly. As a benefit it's not our problem." "But as a hazard," Rachael replies, "then you come in." The first images the audience sees in the movie are shots of an eye in extreme close-up and a magnificent panoramic vista of a futuristic city.[41] The eye might be the "eye of power," scrutinizing and governing the vast cityscape. We soon learn, however, that the eye is Leon's, a suspected Nexus-6 android in the process of being tested with the "Voight-Kampff Empathy Test." The apparatus is used to discover if a suspect is an inhuman replicant. *Blade Runner* thus opens with a lie detector test that poses the movie's central question: "What does it mean to be human?"

Throughout the film, the ocular theme serves to rearticulate the central anxiety of the human-machine opposition. The Voight-Kampff apparatus focuses on the eyes of its suspects. Replicant eyes have a subtle red glow. Seeking information about "Morphology, Longevity, Incept dates," two replicants go to Chew's Eye Works. "If only you could see what I've seen with your eyes," says Batty before executing the eye designer. Eyes are "windows to the soul," but who can possess a soul? The film introduces an interesting complication to this human-machine binary opposition when it suggests that because androids have developed empathy and emotions, it is no longer possible to demarcate between humans and machines. Replicants should, therefore, be able to fool the Voight-Kampff machine, and Rachael, another sophisticated replicant, nearly does so. In a scene replete with film noir signifiers, Deckard explains to Rachael that the instrument "measures capillary dilation in the facial area. We know this to be a primary autonomic response, the so-called 'shame' or 'blushing' reaction to a morally shocking stimulus. It can't be controlled voluntarily, as can skin conductivity, respiration, and cardiac rate." He shows her the other instrument, a pencil-beam light: "This records fluctuations of tension within the eye muscles. Simultaneous with the blush phenomenon there generally can be found a small but detectable movement of —" "And these can't be found in androids," Rachael added.[42]

The movie's designer explained that he wanted the Voight-Kampff apparatus to look like "a giant tarantula on a desk lamp." It was a weird idea, he recalled, but it made him "realize that what could give this sophisticated lie detector a definitely threatening air was to suggest that it was alive." He also devised a small rectangular lens on a stalk that focused on the eye. People were more body-conscious about their eyes than any other organ of the body, he explained. This gave the Voight-Kampff machine an intimidating appearance. "I also designed a set of bellows on the side of the device;" he said, "it breathed. Actually, this breathing had a functional aspect, as the machinery was taking air samples of its subject for analysis. When you're nervous you sweat and exude a distinctive airborne chemistry."[43] Syd Mead placed his futuristic lie detector within an august tradition. The machine was alive; it could breathe and smell fear. It possessed agency and was extremely threatening. As if to emphasize the gaze of the "eye of power," a small screen on the side of the instrument showed a close-up of the suspect's pupil. The eye reacted automatically to stimuli, in the manner of a "primary autonomic response" that "can't be controlled voluntarily." "The VK is used primarily by Blade Runners to determine if a suspect is truly human," the original 1982 *Blade Runner* press kit explained, "by measuring the degree of his empathic response through carefully worded questions and statements."[44]

At the heart of the film lies the problematic status of Rachael, the highly evolved replicant femme fatale. She is the spider woman, the dark lady who is central to the film's key theme of what it is to be human. Like a long line of female suspects before her, she is unfathomable, enigmatic, and inscrutable. Possessing a heightened capacity to deceive, Rachael is the ultimate manifestation of the cultural positioning of women as duplicitous. From Eve

The lie detector represented the dreams of criminology in support of the law. But it also promised to replace the due processes of law altogether. Image from "The Simpsons."

to Pandora, it has been suggested, woman is framed as the perennial problem confronting the will to truth in spite of—or indeed because of—their inscrutability.[45] Film noir habitually places the problem of "woman" herself, not merely the solving of a crime, at the heart of the investigative quest effected by the male detective.[46] The enigmatic status of woman has haunted criminology since its inception in the nineteenth century. "Woman" was the puzzle that the lie detector promised to solve. The scene in which Rachael is interrogated with the Voight-Kampff Empathy Test is crucial to the film's narrative, because it reveals that she is unaware of her status as a replicant. The test breaks Rachael's spirit, shattering her confidence and poise. Later on Deckard reveals that the story she invoked as evidence of her humanity—the dream of a swarm of baby spiders that consume their own mother—was nothing but a false memory, a factory-set implantation. Rachael is crushed by the disclosure.

A central irony of *Blade Runner* is that one apparently sentient machine is used to test the vital integrity of another, further critiquing the apparent human-machine opposition. The term "humanoid android" suggests that the distinction is problematic from the outset. In the film's final scene, the replicants' charismatic leader, Roy Batty, commits an act of such moral commitment that through his actions he has become indistinguishable from a human being he so wants to become. Deckard, the Blade Runner bounty hunter, it emerges in a dramatic twist, might not be human either, but also a replicant.[47] Ultimately the film suggests that what defines the human is the possession neither of memories (for these can be implanted); nor emotions (for these can be acquired); nor even self-knowledge (for this is gained through agency): it is the capacity for ethical action. In this sense, the film concludes, the replicants have indeed become human through their acquisition of a capacity for empathy and self-sacrifice. The film's message is that being human is a matter of ethical action, not genetic inheritance.

The lie detector also raised questions concerning the demarcation between the human and the machine—a consequence of the network of binary oppositions that made its emergence possible. It was the very essence of sober science, but it was a prized resource for entertainers, advertisers, and utopian visionaries. It was a humane technology of truth, although it sought confessions through intimidation. It represented the dreams of a criminology in support of the law, but it promised to replace the due processes of law altogether. It offered to explore the deep recesses of the body yet operated through a veneer of signs. The human subject was construed as possessing mechanistic autonomic responses, but the machine was attributed with humane agency. While the lie detector enjoyed autonomy and charisma, the suspect was regarded as an anonymous automaton. Machines, like dreams, can "fluctuate between being a benefit and a hazard very rapidly." But the problem was not so much deciding when the lie detector was beneficial to humanity and when it was hazardous. The problem was deciding where the machine ended and the human began.

ACKNOWLEDGMENTS

I would like to thank the Bancroft Library, University of California, Berkeley, for permission to quote from the August Vollmer Papers; the Dibner Collection at the Smithsonian Institution, Washington, D.C., for permission to quote from the William Moulton Marston papers; and the Archives of the History of American Psychology, University of Akron, for permission to quote from the Boder Museum Papers.

I can trace the origins of this book to a stimulating period I spent with Geoffrey Cantor, John Christie, Jon Hodge, and Bob Olby at the Centre for History and Philosophy of Science at the University of Leeds. At York University, Toronto, and at the University of Toronto, I was privileged to be able to study with some outstanding scholars including Ray Fancher, Paul Fayter, Chris Green, Ian Hacking, Trevor Levere, Bernie Lightman, and Mariana Valverde. My Ph.D. dissertation supervisor, Kurt Danziger, was, and remains, a great inspiration. My fellow graduate students made my time in Toronto both intellectually invigorating and great fun. I am grateful to my cousin Stacey Crinson and her family for looking after me while I lived in Canada. Ben Harris was an early champion of my work and has continued to send me newspaper and magazine cuttings ever since.

I am grateful to David Borwick, Geoff Bunn Sr., Erica Burman, Hugh Hornby, Mark Jepson, and Graham Richards, all of whom provided insightful comments on earlier drafts of the manuscript. Thanks are also due to Steve and Wil Bunn for helping with the production of the initial book proposal. At the Johns Hopkins University Press I have been fortunate to work with Robert J. Brugger, whose timely interventions have been critically important for the success of this book; and Helen Myers, whose patient copy editing greatly improved the text. My wife, Janet Bunn, has been a perceptive editor and critic. Finally, this book would not have been possible without the love and support of my parents. I dedicate this book to the memory of my mother, Florence Bunn.

NOTES

Introduction. Plotting the Hyperbola of Deception

1. "Lie Test Shows O. J. Didn't Do It!" *The Globe*, February 7, 1995, 5.
2. Mark Nykanen, director, "OJ's Voice Stress Test," *Hard Copy*, January 30, 1995.
3. "Lie Test Shows O.J. Didn't Do It!"
4. See, for example, Paul V. Trovillo, "A History of Lie Detection," pts. 1 and 2, *Journal of Criminal Law and Criminology* 29 (1939): 848–81; 30 (1939): 104–19; Eugene B. Block, *Lie Detectors: Their History and Use* (New York: David McKay Co., 1977).
5. David T. Lykken, *A Tremor in the Blood: Uses and Abuses of the Lie Detector* (New York: McGraw-Hill, 1981), 2; F. Allen Hanson, *Testing, Testing: Social Consequences of the Examined Life* (Berkeley: University of California Press, 1993).
6. Ibid.
7. Evidence suggests that this number has increased since 1988, even though in that year the U.S. government banned use of the lie detector for private preemployment screening, exempting itself, a public employer, from the ruling.
8. Thorn Bacon, "The Man Who Reads Nature's Secrets," *National Wildlife* 7 (February–March 1969), 4–8.
9. Ibid., 7.
10. For the contemporary status of polygraphy, see Anthony Gale, ed., *The Polygraph Test: Lies, Truth and Science* (London: Sage, 1988); Gershon Ben-Shakhar and John J. Furedy, *Theories and Applications in the Detection of Deception: A Psychophysiological and International Perspective* (New York: Springer-Verlag, 1990).
11. Ken Alder, *The Lie Detectors: The History of an American Obsession* (New York: Free Press, 2007).

Chapter 1. "A thieves' quarter, a devil's den": The Birth of Criminal Man

Epigraph. J. B. Thomson, "The Hereditary Nature of Crime," *Journal of Mental Science* 15 (1870): 489.

1. Malcolm Gaskill, *Crime and Mentalities in Early Modern England* (Cambridge: Cambridge University Press, 2000), 203.

2. Ibid., 217.

3. Ibid., 229.

4. Brian Marriner, *Forensic Clues to Murder: Forensic Science in the Art of Crime Detection* (London: Arrow, 1991), 162.

5. "Cruentation (from *cruentare*: to make bloody, to spot with blood) was a test used to find a murderer. Bleeding was considered a "Judgment of God," manifested by the "indignation" of the corpse when the murderer was in its presence. Dating from the period following the overthrow of the Roman Empire, it was used in Europe until at least the seventeenth century. See Robert P. Brittain, "Cruentation in Legal Medicine and in Literature," *Medical History* 9, no. 1 (1965): 82.

6. David Garland, "Of Crimes and Criminals: The Development of Criminology in Britain," in Mike Maguire, Rod Morgan, and Robert Reiner, *The Oxford Handbook of Criminology*, 2nd ed. (Oxford: Oxford University Press, 2007), 22.

7. Ibid., 25.

8. Nicole Hahn Rafter, "The Unrepentant Horse-Slasher: Moral Insanity and the Origins of Criminology," *Criminology* 42 (2004): 979–1008.

9. Sir George Onesiphorus Paul (1809) quoted in Martin J. Wiener, *Reconstructing the Criminal: Culture, Law, and Policy in England, 1830–1914* (Cambridge: Cambridge University Press, 1990), 104.

10. Richard F. Wetzell, *Inventing the Criminal: A History of German Criminology, 1880–1945* (Chapel Hill: University of North Carolina Press, 2000), 32.

11. Wiener, *Reconstructing the Criminal*, 100.

12. Ibid., 103.

13. Michel Foucault, "The Dangerous Individual," in *Politics, Philosophy, Culture: Interviews and Other Writings 1977–1984*, ed. Lawrence D. Kritzman (New York: Routledge, 1988), 127–28.

14. Marie-Christine Leps, *Apprehending the Criminal: The Production of Deviance in Nineteenth-Century Discourse* (Durham, NC: Duke University Press, 1992).

15. Rick Rylance, *Victorian Psychology and British Culture 1850–1880* (Oxford: Oxford University Press, 2000).

16. Wiener, *Reconstructing the Criminal*, 162.

17. Michael Hagner, "Skulls, Brains, and Memorial Culture: On Cerebral Biographies of Scientists in the Nineteenth Century," *Science in Context* 16 (2003): 195–218.

18. See for example, "R," "Social and Moral Statistics of Criminal Offenders," *Journal of the Statistical Society of London* 2, no. 6 (January 1840): 442–45; Theodore M. Porter, *The Rise of Statistical Thinking, 1820–1900* (Princeton: Princeton University Press, 1986).

19. Garland, "Of Crimes and Criminals," 26.

20. Quoted in Peter J. Hutchings, *The Criminal Spectre in Law, Literature and Aesthetics* (London: Routledge, 2001), 172.

21. Quetelet (1835) quoted in Wiener, *Reconstructing the Criminal*, 163.

22. Piers Beirne, "Adolphe Quetelet and the Origins of Positivist Criminology," *American Journal of Sociology* 92, no. 5 (1987): 1160.

23. Quoted in Beirne, "Adolphe Quetelet," 1163.

24. Ian Hacking, "Biopower and the Avalanche of Numbers," *Humanities and Society* 5 (1983): 279–95.

25. Mayhew (1856) cited in Daniel Pick, *Faces of Degeneration: A European Disorder, c. 1848–1918* (Cambridge: Cambridge University Press, 1989), 183.

26. Mayhew (1851) quoted in Wiener, *Reconstructing the Criminal*, 31.

27. B. A. Morel quoted in Kelly Hurley, *The Gothic Body: Sexuality, Materialism, and Degeneration at the Fin-de-Siècle* (Cambridge: Cambridge University Press, 1996), 66, citing Max Simon Nordau, *Degeneration*, trans. from 2nd German ed. (London: Heinemann, 1895), 16.

28. Wetzell, *Inventing the Criminal*, 19–20; Rafter, "The Unrepentant Horse-Slasher."

29. Rafter, "The Unrepentant Horse-Slasher," 1002, 991.

30. Wiener, *Reconstructing the Criminal*, 229.

31. Ibid., 338.

32. Ibid., 166.

33. John Van Wyhe, *Phrenology and the Origins of Victorian Scientific Naturalism* (Aldershot: Ashgate Publishing, 2004); Nicole Hahn Rafter, "The Murderous Dutch Fiddler: Criminology, History and the Problem of Phrenology," *Theoretical Criminology* 9, no. 1 (2005): 65–96.

34. Rafter, "The Murderous Dutch Fiddler," 65, 66.

35. Hewett Watson (1836) quoted in David de Giustino, *Conquest of Mind: Phrenology and Victorian Social Thought* (London: Croom Helm, 1975), 146.

36. On "technologies of the self" see Nikolas Rose, *Inventing Our Selves: Psychology, Power, and Personhood* (Cambridge: Cambridge University Press, 1996).

37. Some examples are: James Simpson, *The Necessity of Popular Education, as a National Object; with Hints on the Treatment of Criminals and Observations of Homicidal Insanity* (Edinburgh: Adam and Charles Black, 1834); George Combe, *Remarks on the Principles of Criminal Legislation and the Practice of Prison Discipline* (London: Simpkin, Marshall and Co., 1854); Marmaduke B. Sampson, *Rationale of Crime, and its Appropriate Treatment: Being a Treatise on Criminal Jurisprudence Considered in Relation to Cerebral Organization* (New York: D. Appleton, 1846); James P. Browne, *Phrenology and its Application to Education, Insanity and Prison Discipline* (London: Bickers and Son, 1869).

38. Quoted in Hagner, "Skulls, Brains, and Memorial Culture," 200.

39. A Member of the Phrenological and Philosophical Societies of Glasgow, *The Philosophy of Phrenology Simplified* (Glasgow: W. R. McPhun, 1838), 185–86.

40. A Member, *The Philosophy of Phrenology Simplified*, 192.

41. Thomas Stone, *Observations on the Phrenological Development of Burke, Hare and Other Atrocious Murderers* (Edinburgh: Robert Buchanan, 1829), 13.

42. For a recent account of this case see Zbigniew Kotowicz, "The Strange Case of Phineas Gage," *History of the Human Sciences* 20 (2007): 115–31.

43. Quoted in F. G. Barker, "Phineas among the Phrenologists: The American Crowbar Case and Nineteenth-Century Theories of Cerebral Localization, *Journal of Neurosurgery* 82 (1995): 678.

44. Wetzell, *Inventing the Criminal*, 17–18.

45. James De Ville, *Manual of Phrenology as an Accompaniment to the Phrenological Bust* (London, 1828), 31.

46. De Ville, *Manual of Phrenology*, 32.

47. Combe, *Remarks on the Principle of Criminal Legislation*, 36.

48. Ibid., 37

49. Frederick Bridges, *Criminals, Crimes, and their Governing Laws, as Demonstrated by the Sciences of Physiology and Mental Geometry* (London: George, Philip and Son, 1860), preface, n.p.

50. Ibid., 8.

51. Ibid., 18.

52. Ibid., 22.

53. Ibid., 23.

54. Combe (1841) quoted in Rafter, "The Murderous Dutch Fiddler," 77.

55. Rafter, "The Murderous Dutch Fiddler," 79.

56. de Giustino, *Conquest of Mind*, chap. 7.

57. Roger Cooter, *The Cultural Meaning of Popular Science: Phrenology and the Organization of Consent in Nineteenth-Century Britain* (Cambridge: Cambridge University Press, 1984).

58. Rafter, "The Murderous Dutch Fiddler," 75.

59. Not everyone read benevolence and reform into phrenology, however: "Because we should be able to identify on a person's skull the marks of serious villainy, the state should prescribe an examination of the skull for everyone who reaches the age of twenty-five. Everyone found guilty of having a dangerous predisposition should be hanged or confined preventively, depending on his anticipated offense!" Gustav Zimmerman (1845) quoted in Peter Becker and Richard F. Wetzell, *Criminals and their Scientists: The History of Criminology in International Perspective* (Cambridge: Cambridge University Press, 2006), 8.

60. On the importance of character in the nineteenth century see Warren I. Susman, "Personality and the Making of Twentieth-century Culture," in *Culture as History: The Transformation of American Society in the Twentieth Century* (New York: Pantheon, 1984), 271–85; Stefan Collini, "The Idea of 'Character' in Victorian Political Thought," *Transactions of the Royal Historical Society, Fifth Series* 35 (1985): 29–50; Melanie White and Alan Hunt, "Citizenship: Care of the Self, Character and Personality," *Citizenship Studies* 4, no. 2 (2000): 93–116; Ben Weinstein, "'Local Self-Government Is True Socialism': Joshua Toulmin Smith, the State and Character Formation," *English Historical Review* 123, no. 504 (2008): 1193–1228.

61. Wiener, *Reconstructing the Criminal*, 45.

62. Ibid., 91.

63. Charles Bray, "The Physiology of the Brain," *Anthropological Review* 7, no. 26 (1869): 271.

64. Ibid., 275.

65. Ibid., 277. On the phrenologists' organized opposition to the transportation of convicts, see de Giustino, *Conquest of Mind*, 153–62.

66. Weinstein, "Local Self-Government Is True Socialism," 1199.

67. T. S. Clouston, "The Developmental Aspects of Criminal Anthropology," *The Journal of the Anthropological Institute of Great Britain and Ireland* 23 (1894): 216.

68. Girard de Rialle, "French Anthropology," *The Journal of the Anthropological Institute of Great Britain and Ireland* 9 (1880): 234.

69. Ibid.

70. J. B. Thomson, "The Hereditary Nature of Crime," *Journal of Mental Science* 15 (1870): 488 (emphasis in original).

71. Ibid., 491.

72. Ibid., 497–98.

73. Ibid., 489.

74. Ibid., 490.

75. Ibid., 494.

76. Ibid., 498.

77. Lombroso in Mary S. Gibson, "Cesare Lombroso and Italian Criminology: Theory and Politics," in Becker and Wetzell, *Criminals and their Scientists,* 139.

78. Thomson, "The Hereditary Nature of Crime," 487. See also J. B. Thomson, "The Psychology of Criminals," *Journal of Mental Science* 17 (1870): 321–50.

79. Thomson, "The Hereditary Nature of Crime," 489.

80. Ibid., 496.

81. Ibid., 498.

82. Wiener, *Reconstructing the Criminal,* 35.

83. Henry Maudsley, *Responsibility in Mental Disease,* 5th ed. (London: C. Kegan Paul and Co., 1892), 28, quoted in Wiener, *Reconstructing the Criminal,* 232.

84. Henry Maudsley, *Body and Mind* (London: Macmillan, 1873), 135.

85. Maudsley, *Responsibility in Mental Disease,* 29–30. On Maudsley see Pick, *Faces of Degeneration,* 203–16, and Elaine Showalter, *The Female Malady: Women, Madness and English Culture, 1830–1980* (London: Virago Press, 1987).

86. Maudsley, *Responsibility in Mental Disease,* 22.

87. Rafter, "The Murderous Dutch Fiddler."

88. Ibid.

89. I owe this charismatic anecdote and much of what follows to Mary Gibson, *Born to Crime: Cesare Lombroso and the Origins of Biological Criminology* (Westport, CT: Praeger, 2002), 9.

90. Ottolenghi (1908) quoted in Gibson, *Born to Crime,* 135.

91. Lombroso (1879) quoted in Gibson, *Born to Crime,* 135.

92. Gibson, *Born to Crime,* 135.

93. Ottolenghi (1914) quoted in Gibson, *Born to Crime,* 138.

94. Cesare Lombroso, *Criminal Man According to the Classification of Cesare Lombroso, Briefly Summarized by his Daughter Gina Lombroso Ferrero, with an Introduction by Cesare Lombroso* (New York: G. P. Putnam's Sons, 1911), xiv.

95. Gibson, *Born to Crime,* 27.

96. Gibson, "Cesare Lombroso and Italian Criminology."

97. David G. Horn, *The Criminal Body: Lombroso and the Anatomy of Deviance* (London: Routledge, 2003).

98. Gibson, "Cesare Lombroso and Italian Criminology," 151.

99. Marvin E. Wolfgang, "Cesare Lombroso," in *Pioneers in Criminology*, ed. Hermann Mannheim, 2nd ed. (Montclair: Patterson Smith, 1972), 250.

100. Wolfgang, "Cesare Lombroso," 251.

101. Gabriel Tarde, *Criminalité Comparée* (Paris: F. Alcan, 1886) quoted in Wolfgang, "Cesare Lombroso," 280.

102. Cited in Gibson, *Born to Crime*, 35.

103. Raffaello Garofalo, *Criminology* (Boston: Little, Brown, 1914), 95, cited in Gibson, *Born to Crime*, 36.

104. Gibson, *Born to Crime*, 35–36.

105. Nicole Hahn Rafter, *Creating Born Criminals* (Champaign: University of Illinois Press, 1997), 128.

106. Clouston, "The Developmental Aspects of Criminal Anthropology," 218 (emphasis added).

107. Ibid., 219.

108. Ibid., 225.

109. Garland, "Of Crimes and Criminals," 28.

110. Richard Bach Jensen, "Criminal Anthropology and Anarchist Terrorism in Spain and Italy," *Mediterranean Historical Review* 16, no. 2 (December 2001): 31–44 (quote on 36).

111. Robert Nye, "Heredity or Milieu: The Foundations of European Criminological Theory," *Isis* 67 (1976): 335–55.

112. But see Martin S. Staum, *Labeling People: French Scholars on Society, Race and Empire, 1815–1848* (Montreal: McGill-Queen's University Press, 2003), 167.

113. On the role that the Congresses played in the emergence of criminology see Martine Kaluszynski, "The International Congresses of Criminal Anthropology: Shaping the French and International Criminological Movement, 1886–1914," in Peter Becker and Richard F. Wetzell, *Criminals and their Scientists: The History of Criminology in International Perspective* (Cambridge: Cambridge University Press, 2006).

114. Lacassagne (1885) cited in Pick, *Faces of Degeneration*, 109.

115. Gibson, *Born to Crime*, 42.

116. On the biometric school see Donald A. Mackenzie, *Statistics in Britain, 1865–1930: The Social Construction of Scientific Knowledge* (Edinburgh: Edinburgh University Press, 1981); and Michael Cowles, *Statistics in Psychology: An Historical Perspective* (Mahway: LEA, 2001).

117. On Bertillon's anthropometry see Allan Sekula, "The Body and the Archive," *October* 39 (1986), 3–64; and Hutchings, *The Criminal Spectre*, chap. 5.

118. On Tarde see Christian Borch, "Urban Imitations: Tarde's Sociology Revisited," *Theory, Culture and Society* 22, no. 3 (2005): 81–100.

119. Nye, "Heredity or Milieu," 348.

120. Pick, *Faces of Degeneration*, 140.

121. Staum, *Labeling People*, 167.

122. Nye, "Heredity or Milieu," 346.

123. Gibson, *Born to Crime*.

124. Jensen, "Criminal Anthropology," 36.

125. Ibid., 37.

126. Lombroso, *Gli Anarchici* (1894), 21, quoted by Jensen, "Criminal Anthropology," 33.
127. Jensen, "Criminal Anthropology," 35.
128. Ibid., 40.
129. Ibid.
130. Wetzell, *Inventing the Criminal*, 39.
131. Koch (1894) in Wetzell, *Inventing the Criminal*, 54.
132. Wetzell, *Inventing the Criminal*, 47.
133. Ibid., 297–98.
134. Roland Grassberger, "Hans Gross," 1847–1915, in *Pioneers of Criminology*, ed. H. Mannheim, 2nd ed. (Montclair, NJ: Patterson Smith, 1972), 305–17 (quote on 307).
135. Lombroso-Ferrero (1911), 135, quoted in Rafter, *Creating Born Criminals*, 106.
136. Lombroso (1876) quoted in Mary Gibson and Nicole Hahn Rafter, *Criminal Man by Cesare Lombroso* (Durham, NC: Duke University Press, 2006), 24.
137. Wetzell, *Inventing the Criminal*, 17.
138. Although all ended up as ingredients in the modern criminological mixture, at the time "they were discrete forms of knowledge, undertaken for a variety of different purposes, and forming elements within a variety of different discourses, none of which corresponded exactly with the criminological project that was subsequently formed." Garland, "Of Crimes and Criminals," 28.
139. Ibid.
140. Wiener, *Reconstructing the Criminal*, 15.

Chapter 2. "A vast plain under a flaming sky": The Emergence of Criminology

Epigraph. Jarkko Jalava, "The Modern Degenerate: Nineteenth-century Degeneration Theory and Psychopathy Research," *Theory and Psychology* 16 (2006): 419.
1. Marie-Christine Leps, *Apprehending the Criminal: The Production of Deviance in Nineteenth-Century Discourse* (Durham, NC: Duke University Press, 1992), 67.
2. Daniel Pick, *Faces of Degeneration: A European Disorder, c. 1848–1918* (Cambridge: Cambridge University Press, 1989), 7; Mary Gibson, *Born to Crime: Cesare Lombroso and the Origins of Biological Criminology* (Westport, CT: Praeger, 2002), 7.
3. Robert Nye, "Heredity or Milieu: The Foundations of European Criminological Theory," *Isis* 67 (1976): 336.
4. Martin J. Wiener, *Reconstructing the Criminal: Culture, Law, and Policy in England, 1830–1914* (Cambridge: Cambridge University Press, 1990), 57; David Garland, "Of Crimes and Criminals: The Development of Criminology in Britain," in Mike Maguire, Rod Morgan, and Robert Reiner, *The Oxford Handbook of Criminology*, 2nd ed. (Oxford: Oxford University Press, 2007), 32–33.
5. Leps, *Apprehending the Criminal*, chap. 3.
6. Garland, "Of Crimes and Criminals," 33.
7. Wiener, *Reconstructing the Criminal*, 226.
8. Garland, "Of Crimes and Criminals," 12.
9. Peter J. Hutchings, *The Criminal Spectre in Law, Literature and Aesthetics* (London: Routledge, 2001), 185.

10. David G. Horn, *The Criminal Body: Lombroso and the Anatomy of Deviance* (London: Routledge, 2003), 29.
11. Pick, *Faces of Degeneration*, 118.
12. Pick, *Faces of Degeneration*, 111; Gibson, *Born to Crime*, 2.
13. Mary Gibson, *Born to Crime*, 3.
14. Lombroso (1871) quoted in Pick, *Faces of Degeneration*, 126.
15. Pick, *Faces of Degeneration*, 128.
16. Lombroso (1894) quoted in Pick, *Faces of Degeneration*, 119.
17. Gibson, *Born to Crime*, chap. 1.
18. Pick, *Faces of Degeneration*, 113.
19. Gibson, *Born to Crime*.
20. Pick, *Faces of Degeneration*, 136.
21. Horn, *The Criminal Body*, chap. 6.
22. Nye, "Heredity or Milieu," 345.
23. Gibson, *Born to Crime*, 6.
24. Gibson, *Born to Crime*, 129.
25. Neil Davie, *Tracing the Criminal: The Rise of Scientific Criminology in Britain, 1860–1918* (Oxford: Bardwell Press, 2006).
26. Pick, *Faces of Degeneration*, 178.
27. Wiener, *Reconstructing the Criminal*, 16.
28. Ibid., 216–17.
29. Pick, *Faces of Degeneration*, 182.
30. Wiener, *Reconstructing the Criminal*, 224.
31. Pick, *Faces of Degeneration*, 150.
32. Tarde (1886) in Leps, *Apprehending the Criminal*, 51.
33. Leps, *Apprehending the Criminal*, chap. 3.
34. T. S. Clouston, "The Developmental Aspects of Criminal Anthropology," *The Journal of the Anthropological Institute of Great Britain and Ireland* 23 (1894): 221 (emphasis in original).
35. Clouston (1906) quoted in Roger Smith, "'Inhibition' and the Discourse of Order," *Science in Context* 5, no. 2 (1992): 237–63.
36. Smith, "'Inhibition' and the Discourse of Order," 248.
37. Wiener, *Reconstructing the Criminal*, 11–12.
38. Mayhew (1862) quoted in Wiener, *Reconstructing the Criminal*, 24–25.
39. Wiener, *Reconstructing the Criminal*, 233.
40. Wiener, *Reconstructing the Criminal*, 234. For an excellent account of the troubled workings of a mid-Victorian local prison that examines prison discipline beyond the Benthamite vision, see Richard W. Ireland, *"A Want of Order and Good Discipline": Rules, Discretion and the Victorian Prison* (Cardiff: University of Wales Press, 2007).
41. Garland, "Of Crimes and Criminals," 36.
42. Garland, "Of Crimes and Criminals," 35.
43. "The Romance of Crime, Criminal and Police," *The Belfast News-Letter* (Belfast), Issue 26020, Tuesday, December 27, 1898, 5.
44. Wiener, *Reconstructing the Criminal*, 217.

45. Lombroso clearly was a household name in Britain, contrary to what Pick, *Faces of Degeneration,* 180, claims. His ideas were regularly discussed—often skeptically—in metropolitan and provincial papers such as *The Pall Mall Gazette,* The London *Graphic, The Glasgow Herald,* the *Aberdeen Weekly Journal,* and the *Hampshire Telegraph and Sussex Chronicle.* See Helen Zimmern, "Professor Lombroso's New Theory of Political Crime," *Blackwood's Magazine* 49 (1891): 202–11; Isabel Foard, "The Criminal: Is He Produced by Environment or Atavism?" *Westminster Review* 150 (1898): 90–103.

46. Davie, *Tracing the Criminal,* chap. 5.

47. Pick, *Faces of Degeneration,* 189–203.

48. Ellis (1890) quoted in Pick, *Faces of Degeneration,* 178.

49. Havelock Ellis, "Retrospect of Criminal Anthropology," *Journal of Mental Science* 37 (1891): 299–309, 458–64.

50. "British Medical Congress," *The Bristol Mercury and Daily Post* (Bristol), Issue 14425, Saturday, August 4, 1894, 5.

51. Havelock Ellis, "Retrospect of Criminal Anthropology," 459. An important figure in the dissemination of criminal anthropology in Britain, Morrison wrote the introduction to the English translation of Cesare Lombroso and Guglielmo Ferrero, *The Female Offender* (New York: D. Appleton, 1895). See also L. Gordon Rylands, *Crime: Its Causes and Remedy* (London: T. Fisher Unwin, 1889).

52. Quoted in Pick, *Faces of Degeneration,* 183.

53. Major Arthur Griffiths, *Mysteries of Crime and Police,* 2 vols. (London: Cassell, 1899); Griffiths, *Fifty Years of Public Service* (London: Cassell, 1904).

54. Quoted in Davie, *Tracing the Criminal,* 234.

55. Ibid., 235.

56. Darwin (1914–15) quoted in Wiener, *Reconstructing the Criminal,* 358.

57. Garland, "Of Crimes and Criminals," 41.

58. Nicole Hahn Rafter, *Creating Born Criminals* (Champaign: University of Illinois Press, 1997), 6.

59. Lombroso-Ferrero, xxix, in Nicole Hahn Rafter, "Criminal Anthropology: Its Reception in the United States and the Nature of Its Appeal," in Peter Becker and Richard F. Wetzell, *Criminals and their Scientists: The History of Criminology in International Perspective* (Cambridge: Cambridge University Press, 2006), 180–81.

60. Rafter, *Creating Born Criminals,* 125.

61. Rafter, "Criminal Anthropology."

62. Rafter, *Creating Born Criminals,* 115.

63. Joseph Jastrow, "A Theory of Criminality," *Science* 8, no. 178 (July 2, 1886): 20–22.

64. Ibid., 22.

65. Robert Fletcher, "The New School of Criminal Anthropology," *American Anthropologist* 4, no. 3 (1891): 201–36.

66. Ibid., 206.

67. See Havelock Ellis and Alexander Winter, *The New York State Reformatory in Elmira* (London: S. Sonnenschein, 1891); Francis J. Lane, *Twelve Years in a Reformatory: A Report of the Activities and Experiences of a Catholic Chaplain During Twelve Years' Service in the El-*

mira Reformatory (New York: The Elmira Reformatory, 1934). See also H. S. Williams, "Can the Criminal Be Reclaimed?" *North American Review* 163, no. 2 (1896): 207–18, who cites the reforms of Elmira as evidence that challenges criminal anthropology.

68. Fletcher, "The New School of Criminal Anthropology," 232.

69. D. G. Brinton, "Current Notes on Anthropology.—V: Criminal Anthropology," *Science* 19, no. 483 (May 6, 1892): 255.

70. D. G. Brinton. "Current Notes on Anthropology.—XLI: The So-called 'Criminal Type'" *Science* 23, no. 579 (March 9, 1894): 127.

71. Brinton, "Current Notes on Anthropology.—XLI," 127.

72. Rafter, *Creating Born Criminals*, 117–18. In all, Rafter has identified nine authors—three social welfare workers, three educators, and three ministers—whom she considers to be the main protagonists of "the concept of the criminal as a physically distinct, atavistic human being."

73. Rafter, *Creating Born Criminals*, 120.

74. Lydston (1904) quoted in Rafter, *Creating Born Criminals*, 124.

75. McKim (1900) quoted in Rafter, *Creating Born Criminals*, 124 (emphases in original).

76. Rafter, "Criminal Anthropology," 166.

77. Horn, *The Criminal Body*, 133–34.

78. Mary Gibson, "Science and Narrative in Italian Criminology, 1880–1920," in *Crime and Culture: An Historical Perspective*, ed. Amy Gilman Srebnick and René Lévy (Aldershot: Ashgate, 2005), 37–47. Quote on 39.

79. Jarkko Jalava, "The Modern Degenerate: Nineteenth-century Degeneration Theory and Psychopathy Research," *Theory and Psychology* 16 (2006): 419.

80. Gibson, "Science and Narrative in Italian Criminology," 38.

81. Ibid., 40.

82. Quoted in Rafter, *Creating Born Criminals*, 112.

83. Gibson, "Science and Narrative in Italian Criminology," 42.

84. Lombroso and Ferrero, *The Female Offender*, 109–10.

85. Kelly Hurley, *The Gothic Body: Sexuality, Materialism, and Degeneration at the* Fin-de-Siècle (Cambridge: Cambridge University Press, 1996), 95.

86. Lombroso and Ferrero, *The Female Offender*, 72–73.

87. Ibid., 129.

88. Ibid.,148–58.

89. Ibid., 131–32.

90. Gibson, "Science and Narrative in Italian Criminology."

91. Ibid., 40–47.

92. Ibid., 47.

93. Bela Földes, "The Criminal," *Journal of the Royal Statistical Society* 69, no. 3 (September 1906): 567.

94. "An Epidemic of Kissing in America: A Novel Subject Treated from an Entirely New Point of View by Professor Lombroso, the Famous Italian Psychologist," *The Pall Mall Gazette* 10707, July 21, 1899, 10.

95. D. G. Brinton, "Current Notes on Anthropology.—VII: The Criminal Anthropology of Woman," *Science* 19, no. 487 (June 3, 1892): 316.

96. Adalbert Albrecht, "Cesare Lombroso: A Glance at His Life Work," *Journal of the American Institute of Criminal Law and Criminology* 1, no. 2 (July 1910): 73.

97. Havelock Ellis, *The Criminal*, 5th ed. (London: Walter Scott Publishing Co., 1914), 250.

98. Hurley, *The Gothic Body*, 93.

99. Marvin E. Wolfgang, "Cesare Lombroso, 1835–1909," in *Pioneers in Criminology*, ed. Hermann Mannheim, 2nd ed. (Montclair, NJ: Patterson Smith Publishing, 1972), 262.

100. Quoted in Jalava, "The Modern Degenerate," 418.

101. Rafter, "Criminal Anthropology."

102. Leps, *Apprehending the Criminal*, 44.

103. Ibid., 48.

104. Rafter, *Creating Born Criminals*, 113.

105. Quoted in Wolfgang, "Cesare Lombroso, 1835–1909," 261.

106. Gustave Tarde, "Is There a Criminal Type?" *Charities Review* 6, no. 2 (April 1897): 112. Gustave Tarde was at the Bureau of Statistics at the Ministry of Justice in Paris.

107. Rafter, "Criminal Anthropology," 159.

108. Ibid., 176.

109. Ibid., 178.

110. Garland, "Of Crimes and Criminals," 23.

111. Ottolenghi (1908) quoted in Gibson, *Born to Crime*, 135.

112. Tarde, "Is There a Criminal Type?" 110.

113. Wolfgang, "Cesare Lombroso, 1835–1909," 232.

114. Ibid., 287.

115. Albrecht, "Cesare Lombroso," 72.

116. Alfred Lindesmith and Yale Levin, "The Lombrosoian Myth in Criminology," *The American Journal of Sociology* 42 (1937): 654.

117. Gibson, "Science and Narrative in Italian Criminology," 40.

118. Richard Bach Jensen, "Criminal Anthropology and Anarchist Terrorism in Spain and Italy," *Mediterranean Historical Review* 16, no. 2 (December 2001): 36.

119. Richard F. Wetzell, *Inventing the Criminal: A History of German Criminology, 1880–1945* (Chapel Hill: University of North Carolina Press, 2000), 53.

120. Quoted in Stephen Jay Gould, *The Mismeasure of Man* (London: Penguin Books, 1981), 135.

121. In Pick, *Faces of Degeneration*, 121.

122. Tarde, "Is There a Criminal Type?"

123. Mary S. Gibson, "Cesare Lombroso and Italian Criminology: Theory and Politics," in Becker and Wetzell, *Criminals and their Scientists*, 141.

124. Wetzell, *Inventing the Criminal*, 30.

125. Horn, *The Criminal Body*, 133.

126. Wetzell, *Inventing the Criminal*, 31.

127. Cesare Lombroso, "Atavism and Evolution," *Contemporary Review* 68 (July/December 1895): 42–49.

128. Max Weber, "The Sociology of Charismatic Authority/The Nature of Charismatic Authority and Its Routinization," in *The Celebrity Culture Reader*, ed. P. David Marshall (London: Routledge, 2006), 60.

129. Ibid., 56.
130. Ibid., 61.
131. Charles Thorpe and Steven Shapin, "Who Was J. Robert Oppenheimer?: Charisma and Complex Organization," *Social Studies of Science* 30 (2000): 580.
132. Pick, *Faces of Degeneration*, 149.
133. Quoted in Wetzell, *Inventing the Criminal*, 56–57.
134. Tarde, "Is There a Criminal Type?" 112.
135. Leps, *Apprehending the Criminal*, 220.
136. Ibid.
137. Tarde, "Is There a Criminal Type?" 109.
138. Ibid.
139. Ibid.

Chapter 3. "Supposing that Truth is a woman—what then?": The Enigma of Female Criminality

Epigraphs. Sigmund Freud, "Three Essays on Sexuality," in Peter Gay, ed. *The Freud Reader* (London: Vintage, 1995): 248; Friedrich Nietzsche, *Beyond Good and Evil: Prelude to a Philosophy of the Future*, trans. R. J. Hollingworth (London: Penguin, 1973), 164.

1. Cynthia Eagle Russett, *Sexual Science: The Victorian Construction of Womanhood* (Cambridge, MA: Harvard University Press, 1989).
2. Quoted in Ornella Moscucci, "Hermaphroditism and Sex Difference: The Construction of Gender in Victorian England," in *Science and Sensibility: Gender and Scientific Enquiry, 1780–1945*, ed. Marina Benjamin (Oxford: Basil Blackwell, 1991), 174.
3. Elaine Showalter, *Sexual Anarchy: Gender and Culture at the Fin de Siècle* (London: Bloomsbury, 1991), 129.
4. Frances Power Cobbe, "Criminals, Idiots, Women and Minors," *Fraser's Magazine* 78 (December 1868): 777–94.
5. Beverly Brown, "Women and Crime: The Dark Figures of Criminology," *Economy and Society* 15 (1986): 401.
6. Alison Young, *Imagining Crime: Textual Outlaws and Criminal Conversations* (London: Sage, 1996), 31.
7. Elizabeth V. Spelman, "Woman as Body: Ancient and Contemporary Views," in *Feminist Theory and the Body: A Reader*, ed. Janet Price and Margrit Shildrick (Edinburgh: Edinburgh University Press, 1999), 39.
8. Carroll Smith-Rosenberg, *Disorderly Conduct: Visions of Gender in Victorian America* (New York: Oxford University Press, 1985), 189–90.
9. Carolyn Merchant, *The Death of Nature: Women, Ecology, and the Scientific Revolution* (San Francisco: Harper and Row, 1983).
10. Natalie Zemon Davis, "Gender and Sexual Temperament," in *The Polity Reader in Gender Studies* (Cambridge: Polity Press, 1994), 129–34.
11. Merchant, *The Death of Nature*.
12. Zemon Davis, "Gender and Sexual Temperament," 131.

13. Quoted in Steven Shapin, *A Social History of Truth: Civility and Science in Seventeenth-Century England* (Chicago: University of Chicago Press, 1994), 90.
14. Shapin, *A Social History of Truth*, 83–84.
15. Thomas Laqueur, *Making Sex: Body and Gender from the Greeks to Freud* (Cambridge, MA: Harvard University Press, 1990), 149.
16. Susan J. Hekman, *Gender and Knowledge: Elements of a Postmodern Feminism* (Boston: Northeastern University Press, 1990), 111.
17. Quoted in Eva Figes, *Patriarchal Attitudes: Women in Society* (New York: Persea Books, 1970), 122.
18. Smith-Rosenberg, *Disorderly Conduct*, 13.
19. Ibid., 25.
20. Quoted by Marina Benjamin, "Introduction," in *Science and Sensibility*, ed. Benjamin, 1.
21. Smith-Rosenberg, *Disorderly Conduct*, 183.
22. Karen Lystra, *Searching the Heart: Women, Men, and Romantic Love in Nineteenth-Century America* (Oxford: Oxford University Press, 1989), 20.
23. Darwin quoted in Figes, *Patriarchal Attitudes*, 113–14.
24. Mary Poovey, *Uneven Developments: The Ideological Work of Gender in Mid-Victorian England* (Chicago: University of Chicago Press, 1988), 6.
25. Geneviève Fraisse, "A Philosophical History of Sexual Difference," in *A History of Women in the West: IV, Emerging Feminism from Revolution to World War*, ed. Geneviève Fraisse and Michelle Perrot (Cambridge, MA: Harvard University Press, 1993), 61.
26. George Beard, *American Nervousness: Its Causes and Consequences* (New York: Putnam, 1881), vi.
27. Elaine Showalter, *The Female Malady: Women, Madness and English Culture, 1830–1980* (London: Virago Press, 1987), 122.
28. Quoted in Toril Moi, *What is a Woman? And Other Essays* (Oxford: Oxford University Press, 1999), 18.
29. Showalter, *Female Malady*, 122.
30. Moscucci, "Hermaphroditism and Sex Difference," 193.
31. Dr. M. L. Holbrook (1882) quoted in Poovey, *Uneven Developments*, 35.
32. Smith-Rosenberg, *Disorderly Conduct*, 22–23.
33. Carole Pateman, *The Disorder of Women: Democracy, Feminism and Political Theory* (Cambridge: Polity Press, 1989), 18.
34. Ibid., 76.
35. Quoted in ibid., 76
36. Kant (1764) quoted in Lorraine Daston, "The Naturalized Female Intellect," in *Historical Dimensions of Psychological Discourse*, ed. Carl F. Graumann and Kenneth J. Gergen (Cambridge: Cambridge University Press, 1996), 186.
37. Figes, *Patriarchal Attitudes*, 123.
38. Ibid., 124.
39. Ibid., 125.
40. Nietzsche, *Beyond Good and Evil*, 100.
41. Ibid., 31, 164.

42. On hysteria see Rachel P. Maines, *The Technology of Orgasm: "Hysteria," The Vibrator, and Women's Sexual Satisfaction* (Baltimore: Johns Hopkins University Press, 1999).

43. Mark Micale, *Approaching Hysteria: Disease and its Interpretations* (Princeton: Princeton University Press, 1995), 57.

44. *Physiology, Medicine etc.* (London: Spottiswoode & Co., n.d. [published before 1860])

45. Showalter, *Female Malady*, 145.

46. Jane Ussher, *Women's Madness: Misogyny or Mental Illness?* (New York: Harvester Wheatsheaf, 1991), 91.

47. Showalter, *The Female Malady*, 121–64.

48. Micale, *Approaching Hysteria*, 225.

49. Mark Micale, "Hysteria Male/Hysteria Female: Reflections on Comparative Gender Construction in Nineteenth-Century France and Britain," in *Science and Sensibility*, ed. Benjamin, 200–39. Quote on 205–6.

50. Unattributed, quoted in Juliet Mitchell, *Women: The Longest Revolution: Essays in Feminism, Literature and Psychoanalysis* (London: Virago, 1984), 115.

51. Micale, "Hysteria Male/Hysteria Female," 205–6.

52. Maudsley (1895) quoted in Showalter, *The Female Malady*, 133–34.

53. *Physiology, Medicine etc.*

54. Quoted in Poovey, *Uneven Developments*, 45–46.

55. Ibid., 46.

56. W. L. Distant, "On the Mental Differences Between the Sexes," *The Journal of the Anthropological Institute of Great Britain and Ireland* 4 (1875): 84.

57. Mary Gibson, *Born to Crime: Cesare Lombroso and the Origins of Biological Criminology* (Westport, CT: Praeger, 2002), 65.

58. Lombroso and Ferrero cited in Gibson, *Born to Crime*, 64.

59. Bela Földes, "The Criminal," *Journal of the Royal Statistical Society* 69, no. 3 (September 1906): 559.

60. Maudsley (1874) quoted in Lucia Zedner, *Women, Crime and Custody in Victorian England* (Oxford: Clarendon Press, 1991), 87.

61. Havelock Ellis (1904) quoted in Zedner, *Women, Crime and Custody*, 87.

62. D. G. Brinton, "Current Notes on Anthropology.—VII: The Criminal Anthropology of Woman," *Science* 19, no. 487 (June 3, 1892): 316.

63. Cesare Lombroso and Guglielmo Ferrero, *The Female Offender* (New York: D. Appleton, 1895), 147.

64. David G. Horn, *The Criminal Body: Lombroso and the Anatomy of Deviance* (London: Routledge, 2003), 70.

65. Lombroso and Ferrero, *The Female Offender*, 148.

66. Ibid., 25.

67. Gibson, *Born to Crime*, 61.

68. Ottolengh (1896) cited in Gibson, *Born to Crime*, 62.

69. Lombroso and Ferrero, *The Female Offender*, 111.

70. Lombroso (1892) quoted in Marie-Christine Leps, *Apprehending the Criminal: The Production of Deviance in Nineteenth-Century Discourse* (Durham, NC: Duke University Press, 1992), 62.

71. Brinton, "Current Notes on Anthropology.—VII," 316.

72. A figure confirmed by Földes, "The Criminal," 560, but referring to crime in general across nations.

73. Zedner, *Women, Crime and Custody*, 1.

74. Martin J. Wiener, *Reconstructing the Criminal: Culture, Law, and Policy in England, 1830–1914* (Cambridge: Cambridge University Press, 1990), 130.

75. Gibson, *Born to Crime*, 69. But see Matthew C. Scheider "Moving Past Biological Determinism in Discussions of Women and Crime during the 1870s–1920s: A Note Regarding the Literature," *Deviant Behavior* 21, no. 5 (2000): 407–27.

76. Horn, *The Criminal Body*, 332.

77. Földes, "The Criminal," 558–59.

78. Piers Beirne, "Adolphe Quetelet and the Origins of Positivist Criminology," *American Journal of Sociology* 92, no. 5 (1987): 1157.

79. Földes, "The Criminal," 562.

80. Cited in Gibson, *Born to Crime*, 75.

81. Gibson, *Born to Crime*, 88.

82. Lombroso and Ferrero, *The Female Offender*, 154.

83. Gibson, *Born to Crime*, 88.

84. Adalbert Albrecht, "Cesare Lombroso: A Glance at His Life Work," *Journal of the American Institute of Criminal Law and Criminology* 1, no. 2 (July 1910): 80.

85. Quoted in Kelly Hurley, *The Gothic Body: Sexuality, Materialism, and Degeneration at the Fin-de-Siècle* (Cambridge: Cambridge University Press, 1996), 98.

86. Laqueur, *Making Sex*, 230.

87. Peter J. Hutchings, *The Criminal Spectre in Law, Literature and Aesthetics* (London: Routledge, 2001), 104.

88. Quoted in Hutchings, *The Criminal Spectre*, 106.

89. Quoted in Frances A. Kellor, "Psychological and Environmental Study of Women Criminals I," *The American Journal of Sociology* 5, no. 4 (January 1900): 531.

90. Ibid., 57–58.

91. Ibid., 50.

92. Ibid., 99.

93. Ibid., 100–101.

94. Ibid., 107.

95. Quoted in Marvin E. Wolfgang, "Cesare Lombroso, 1835–1909," in *Pioneers in Criminology*, ed. Hermann Mannheim, 2nd ed. (Montclair: Patterson Smith, 1972), 255.

96. Lombroso and Ferrero, *The Female Offender*, 101.

97. Ibid., 102.

98. Ibid., 107–111.

99. Ibid., 110.

100. Albrecht, "Cesare Lombroso," 77.

101. Havelock Ellis, *The Criminal*, 5th ed. (London: The Walter Scott Publishing Co., 1916), 268.

102. Albrecht, "Cesare Lombroso," 79.

103. Lombroso and Ferrero, *The Female Offender*, 150–51.

104. Albrecht, "Cesare Lombroso," 79.

105. Francis Galton, *Inquiries into Human Faculty and Its Development*, 2nd ed. (London: J. M. Dent and Sons, 1907), 20–21.

106. David Horn, "Making Criminologists: Tools, Techniques, and the Production of Scientific Authority," in *Criminals and their Scientists: The History of Criminology in International Perspective*, ed. Peter Becker and Richard F. Wetzell (Cambridge: Cambridge University Press, 2006), 322–23.

107. Horn, *The Criminal Body*, 89.

108. Cesare Lombroso, "The Physical Insensibility of Women," *Fortnightly Review* n.s. 51 (1892): 354–57.

109. Mary Gibson, "On the Insensitivity of Women: Science and the Woman Question in Liberal Italy, 1890–1910," *Journal of Women's History* 2, no. 2 (1990): 11–41.

110. Gibson, *Born to Crime*, chap. 2.

111. Cited in Hurley, *The Gothic Body*, 98.

112. Lombroso and Ferrero, *The Female Offender*, 151.

113. Thomas M. Dixon, *From Passions to Emotions: The Creation of a Secular Psychological Category* (Cambridge: Cambridge University Press, 2003), 164.

114. Quoted in Hutchings, *The Criminal Spectre*, 102.

115. M. E. Owen, "Criminal Women," *Cornhill Magazine* 14 (August 1866): 152–53.

116. Charles Darwin, *The Expression of the Emotions in Man and Animals* (New York: D. Appleton and Co., 1873), 334.

117. Darwin, *Expression of the Emotions*, 346–47.

118. Horn, "Making Criminologists," 331.

119. Havelock Ellis, *The Criminal*, 138.

120. Darwin, *Expression of the Emotions*, 326.

121. Dixon, *From Passions to Emotion*.

122. Otniel E. Dror, "The Scientific Image of Emotion: Experience and Technologies of Inscription," *Configurations* 7, no. 3 (1999): 357.

123. Anson Rabinbach, *The Human Motor: Energy, Fatigue, and the Origins of Modernity* (New York: Basic, 1990), 96.

124. Dror, "The Scientific Image of Emotion," 358.

125. Ibid.

126. Otniel E. Dror, "Techniques of the Brain and the Paradox of Emotions, 1880–1930," *Science in Context* 14, no. 4 (2001): 646.

127. Horn, *The Criminal Body*, 119.

128. Ibid., 122.

129. Wolfgang, "Cesare Lombroso, 1835–1909," 237.

130. Horn, *The Criminal Body*, 96; Hurley, *The Gothic Body*, 100.

131. Horn, "Making Criminologists," 321.

132. Horn, *The Criminal Body*, 26.

133. Lombroso (1891) quoted in Hurley, *The Gothic Body*, 101.

134. Leps, *Apprehending the Criminal*, 47.

135. Quoted in Horn, *The Criminal Body*, 127.

136. Horn, *The Criminal Body*, 128.

137. J[oseph] J[astrow], "Illustrations of Recent Italian Psychology," *Science* 6, no. 144 (November 6, 1885): 413–15.

138. Lombroso and Ferrero, quoted in Horn, *The Criminal Body*, 126.

139. Horn, *The Criminal Body*, 84.

140. Cited in Horn, *The Criminal Body*, 85.

141. Ibid., 86.

142. Enrico Ferri, *Criminal Sociology*, trans. "W. D. M." (London: T. Fisher Unwin, 1895), 166–67.

143. Gabriel Tarde, *Penal Philosophy*, trans. Rapelje Howell (Boston: Little, Brown and Co., 1912), 63–64.

144. Gina Lombroso-Ferrero, *Criminal Man: According to the Classification of Cesare Lombroso* (Montclair: Patterson Smith, 1911/1972), 223.

145. Ibid., 224–25.

146. Arthur Macdonald, "The Study of Crime and Criminals," *The Chautauquan* 18 (1893): 265–70.

147. Ibid., 268–69.

148. Cesare Lombroso, *Crime: Its Causes and Remedies* (Boston: Little, Brown and Co., 1911), 254.

149. See Helen Zimmern, "Criminal Anthropology in Italy," *Popular Science Monthly* 52, 1897–98, 743–60.

150. Quoted in Hutchings, *The Criminal Spectre*, 107.

151. Ibid., 110.

152. Horn, *The Criminal Body*, 141.

153. Ibid., 87.

154. Young, *Imagining Crime*, 27.

Chapter 4. "Fearful errors lurk in our nuptial couches": The Critique of Criminal Anthropology

Epigraph. Alfred Austin, "Our Novels: The Sensation School," *Temple Bar* 29 (1879): 422.

1. John Kucich, *The Power of Lies: Transgression in Victorian Fiction* (Ithaca: Cornell University Press, 1994).

2. Martin J. Wiener, *Reconstructing the Criminal: Culture, Law, and Policy in England, 1830–1914* (Cambridge: Cambridge University Press, 1990), 245.

3. Ibid., 244.

4. Kate Summerscale, *The Suspicions of Mr. Whicher or The Murder at Road Hill House* (London: Bloomsbury, 2008), xi.

5. "Celebrated Crimes and Criminals—No. XIII," *The Sporting Times* 1248, Saturday, August 20, 1887, 2.

6. Peter J. Hutchings, *The Criminal Spectre in Law, Literature and Aesthetics* (London: Routledge, 2001), 28.

7. Ronald R. Thomas, "The Lie Detector and the Thinking Machine," in *Detective Fiction and the Rise of Forensic Science* (Cambridge: Cambridge University Press, 1999), 35.

8. Wiener, *Reconstructing the Criminal*, 247.

9. Alfred Austin, "Our Novels: The Sensational School," *Temple Bar* 29 (June 1870): 422.
10. Wiener, *Reconstructing the Criminal*, 248.
11. Hutchings, *The Criminal Spectre*, 93.
12. Alison Young, *Imagining Crime: Textual Outlaws and Criminal Conversations* (London: Sage, 1996), 109.
13. Wiener, *Reconstructing the Criminal*, 245.
14. Marie-Christine Leps, *Apprehending the Criminal: The Production of Deviance in Nineteenth-Century Discourse* (Durham, NC: Duke University Press, 1992), 94.
15. Ibid., 99.
16. Ibid., 113.
17. Daniel Pick, *Faces of Degeneration: A European Disorder, c. 1848–1918* (Cambridge: Cambridge University Press, 1989), 4.
18. Ibid., 163.
19. Nils Clausson, "Degeneration, Fin-de-Siècle Gothic, and the Science of Detection: Arthur Conan Doyle's *The Hound of the Baskervilles* and the Emergence of the Modern Detective Story," *Journal of Narrative Theory* 35, no. 1 (2005): 64, 76.
20. Wiener, *Reconstructing the Criminal*, 173.
21. Nicole Hahn Rafter, *Creating Born Criminals* (Champaign: University of Illinois Press, 1997), 38.
22. Kelly Hurley, *The Gothic Body: Sexuality, Materialism, and Degeneration at the Fin-de-Siècle* (Cambridge: Cambridge University Press, 1996), 4.
23. Ibid., 63.
24. Ibid., 4.
25. Ibid., 60.
26. Peter J. Bowler, *The Non-Darwinian Revolution: Reinterpreting a Historical Myth* (Baltimore: Johns Hopkins University Press, 1988).
27. Hurley, *The Gothic Body*, 8.
28. Leps, *Apprehending the Criminal*, 218.
29. Quoted in Pick, *Faces of Degeneration*, 171.
30. Hutchings, *The Criminal Spectre*, 12.
31. Pick, *Faces of Degeneration*, 171.
32. Conan Doyle, "A Scandal in Bohemia" (1891), quoted in Wiener, *Reconstructing the Criminal*, 222.
33. Quoted in Clausson, "Degeneration," 61.
34. Quoted in Clausson, "Degeneration," 74–75, 97 (emphasis added).
35. Ibid., 63.
36. Judith Wilt, "The Imperial Mouth: Imperialism, the Gothic and Science Fiction." *Journal of Popular Culture* 14, no. 4 (1981): 618–28.
37. Wiener, *Reconstructing the Criminal*, 223.
38. See Ronald R. Thomas, "The Fingerprints of the Foreigner: Colonizing the Criminal Body in 1980s Detective Fiction and Criminal Anthropology," *ELH* 61, no. 3 (1994): 655–83.
39. Quoted in Wiener, *Reconstructing the Criminal*, 220.
40. Clausson, "Degeneration," 77.
41. Hutchings, *The Criminal Spectre*, 187.

42. Conan Doyle, "The Final Problem" (1893), quoted in Hutchings, *The Criminal Spectre*, 194.
43. Lombroso in *The Man of Genius* (1864/1891), quoted in Hurley, *The Gothic Body*, 67.
44. Hutchings, *The Criminal Spectre*, 194–95.
45. Quoted in Hurley, *The Gothic Body*, 42.
46. Wiener, *Reconstructing the Criminal*, 251–52.
47. Pick, *Faces of Degeneration*, 158.
48. Hurley, *The Gothic Body*, 104.
49. Ibid., 103.
50. Ibid., 109. Moreau's namesake was Jacques-Joseph Moreau, whose *Morbid Psychology* (1859) posited that the over-excitation of the intellect atrophies the moral sensibility. The book was also an inspiration for Lombroso.
51. Hurley, *The Gothic Body*, 108.
52. Ibid., 113.
53. Marion Shaw, "'To Tell the Truth of Sex': Confession and Abjection in Late Victorian Writing," in *Rewriting the Victorians: Theory, History, and the Politics of Gender*, ed. Linda M. Shires (New York: Routledge, 1992), 92.
54. Wiener, *Reconstructing the Criminal*, 254.
55. Piers Beirne, "Heredity vs Environment: A Reconsideration of Charles Goring's *The English Convict* (1913)," *British Journal of Criminology* 28 (1988): 315–39.
56. W. D. Morrison, "The Study of Crime," *Mind* n.s. 1, no. 4 (October 1892): 489–517.
57. Ibid., 506, 508.
58. "Review of *Criminology* by Arthur MacDonald," *Science* 21, no. 523 (February 10, 1893): 83.
59. Gustave Tarde, "Is There a Criminal Type?," *Charities Review* 6, no. 2 (April 1897): 110.
60. Dr. H. S. Williams, "Can the Criminal Be Reclaimed?," *North American Review* 163, no. 2 (August 1896): 207–18.
61. Ibid., 207.
62. Ibid., 208.
63. Ibid., 210.
64. Ibid., 211.
65. Ibid., 212.
66. Ibid., 213.
67. Ibid., 217.
68. Ibid., 213.
69. Ibid., 216.
70. Frances Alice Kellor, "Sex in Crime," *International Journal of Ethics* 9, no. 1 (October 1898): 74–85.
71. Ibid., 76.
72. Ibid., 81.
73. Ibid., 82.
74. Frances A. Kellor, "Psychological and Environmental Study of Women Criminals I," *The American Journal of Sociology* 5, no. 4 (January 1900): 527–43; Frances A. Kellor, "Psychological and Environmental Study of Women Criminals II," *The American Journal of Sociology* 5, no. 5 (March 1900): 671–82.

75. Kellor, "Psychological and Environmental Study I," 528.
76. Kellor, "Psychological and Environmental Study II," 682.
77. Kellor, "Psychological and Environmental Study I," 529.
78. Ibid., 530.
79. Ibid., 531.
80. Ibid., 532.
81. Ibid., 532.
82. Ibid., 536.
83. Kellor, "Psychological and Environmental Study II," 679 (emphasis in original).
84. Ibid., 681.
85. Kellor, "Psychological and Environmental Study I," 541.
86. Ibid., 541–42.
87. Ibid., 542.
88. Kellor, "Psychological and Environmental Study II," 677.
89. Ibid., 677.
90. It is perhaps significant, given the hitherto male-dominated enterprise of criminology, that Kellor was female. According to Alison Young, fin-de-siècle discourses such as psychoanalysis, criminology, and sexology were convinced that the puzzle of women could be solved thanks to "the suspicion that women are known to themselves, enigmatic only across the divide of sexual difference." Young, *Imagining Crime*, 31.
91. Kellor was the daughter of an impoverished widow. She graduated from Cornell Law School in 1897. A political activist and writer, Kellor founded the National League for the Protection of Colored Women in 1906 and by 1908 was the secretary of the New York State Immigration Commission. She wrote a dozen books and numerous articles, especially in popular periodicals of the progressive era. She was also a founder member and first vice-president of the American Arbitration Association. See "Kellor, Frances Alice (1873–1952)," in Doris Weatherford, *American Women's History: An A to Z of People, Organizations, Issues, and Events* (New York: Prentice Hall, 1994), 195–96; and Lucille O'Connell, "Kellor, Frances," in *Notable American Women: The Modern Period, A Biographical Dictionary*, ed. Barbara Sicherman and Carol Hurd Green (Cambridge, MA: Harvard University Press, 1980), 393–95.

Chapter 5. "To Classify and Analyze Emotional Persons": The Mistake of the Machines

Epigraph. Gilbert K. Chesterton, "The Mistake of the Machines," in *The Complete Father Brown* (New York: Dodd, Mead, and Co., 1914/1982), 298.
1. Frank Marshall White, "The Soul Machine," *Harper's Weekly* 52, December 19, 1908, 12–13, 32.
2. Charles Edmonds Walk, *The Yellow Circle* (New York: A. L. Burt Co., 1909).
3. "Discovery Made by a Swiss Doctor May Play an Important Part in Criminal Trials," *New York Times*, June 9, 1907 V, 8.
4. Ibid.
5. Ibid.

6. Ibid.
7. Ibid.
8. Ibid.
9. Eva Neumann and Richard Blanton, "The Early History of Electrodermal Research," *Psychophysiology* 6, no. 4 (1970): 453–75.
10. E. Prideaux, "The Psychogalvanic Reflex: A Review," *Brain* 43 (1920): 50.
11. Ibid.
12. Frederick Peterson and Carl G. Jung, "Psycho-Physical Investigations with the Galvanometer and Pneumograph in Normal and Insane Individuals," *Brain* 30 (1907): 155; Neumann and Blanton "The Early History of Electrodermal Research," 453–75.
13. Quoted in Paul V. Trovillo, "A History of Lie Detection," *Journal of Criminal Law and Criminology* 30 (1939): 104.
14. Frederick Peterson, "The Galvanometer as a Measurer of Emotions," *The British Medical Journal* (September 28, 1907): 804.
15. Carl G. Jung, "On Psychophysical Relations of the Associative Experiment," *Journal of Abnormal Psychology* 1 (1907): 247.
16. Trovillo reports that Veraguth "was one of the first to make word-association tests with the galvanometer." Trovillo, "A History of Lie Detection," 105.
17. "Discovery Made by a Swiss Doctor" V, 8.
18. For the history of the clinical experiment see Kurt Danziger, *Constructing the Subject: Historical Origins of Psychological Research* (Cambridge: Cambridge University Press, 1990), 52–54.
19. "Emotional complex" was a term Jung had introduced by amalgamating Theodor Ziehen's "gefühlsbetonter vorstellungskomplex" and Pierre Janet's "ideé fixe subconsciente." See Henri F. Ellenberger, *The Discovery of the Unconscious: The History and Evolution of Dynamic Psychiatry* (New York: Basic Books, Inc., 1970), 149, 692–94.
20. "Discovery Made by a Swiss Doctor" V, 8.
21. Peterson, "The Galvanometer as a Measurer of Emotions," 804. See also Frederick Peterson, "The Galvanometer in Psychology," *Journal of Nervous and Mental Disease* 35 (1908): 273–74.
22. Peterson, "The Galvanometer as a Measurer of Emotions," 805. See also Knight Dunlap's critical review "Galvanometric Deflections with Electrodes Applied to the Animal Body," *Psychological Bulletin* 7 (1910): 174–77. "The striking thing about all the work on psychogalvanism is that the enthusiasts have either been unable to conceive of the simplest and most obvious check experiments, or have been unwilling to carry them out" (176).
23. E. G. Boring, *A History of Experimental Psychology* (New York: Appleton-Century-Crofts, Inc., 1957), 546.
24. Boring, *A History of Experimental Psychology*, 527–28.
25. E. W. Scripture, "Detection of the Emotions by the Galvanometer," *Journal of the American Medical Association* (April 11, 1907): 1164 (emphasis added).
26. Jung, "On Psychophysical Relations of the Associative Experiment," 247.
27. Ibid.
28. According to Kurt Danziger, the meaning of term "personality" as used by Jung would

have still suggested a pathological dimension. Personality as a quality of the ordinary (nonpathological) business leader or charismatic movie star had yet to emerge. See Kurt Danziger, *Naming the Mind: How Psychology Found its Language* (London: Sage Publications, 1997), 124–33.

29. Jung, "On Psychophysical Relations of the Associative Experiment," 250.
30. Peterson and Jung, "Psycho-Physical Investigations with the Galvanometer."
31. Charles Ricksher and Carl G. Jung, "Further Investigations on the Galvanic Phenomenon and Respiration in Normal and Insane Individuals," *Journal of Abnormal Psychology* 2 (1907): 203.
32. Ibid., 215.
33. Peterson and Jung, "Psycho-Physical Investigations with the Galvanometer," 172–73.
34. Ibid., 175.
35. "Invents Machines for 'Cure of Liars,'" *New York Times*, September 11, 1907, 9.
36. Ibid.
37. "'I Can Tell if You're a Liar!': Harvard Professor with Strenuous Name Invents Machine that Will Make Him Famous," *New York Times*, September 15, 1907, E8.
38. Hugo Münsterberg, "Traces of Emotion and the Criminal," *Cosmopolitan* 44, April 1908, 528.
39. Hugo Münsterberg, *On the Witness Stand: Essays on Psychology and Crime* (New York: Clark Boardman, 1927), 99–100. First published 1908.
40. "A Scientific Crime Detector," *Scientific American Supplement No. 1666* 64, December 7, 1907, 363.
41. Quoted in Mathew Hale, *Human Science and Social Order: Hugo Münsterberg and the Origins of Applied Psychology* (Philadelphia: Temple University Press, 1980), 119.
42. Jutta Spillmann and Lothar Spillmann, "The Rise and Fall of Hugo Münsterberg," *Journal of the History of the Behavioral Sciences* 29 (1993): 329.
43. See, for example, Hugo Münsterberg, "Hypnotism and Crime," *McClure's Magazine* 30, January 1908, 317–22; and Hugo Münsterberg, "The Prevention of Crime" *McClure's Magazine* 30, April 1908, 750–56.
44. Hugo Münsterberg "Nothing But The Truth," *McClure's Magazine* 29, September 1907, 532–36.
45. "A Psychologist's Judicial Warning," *New York Times*, August 25, 1907 II, 16.
46. Münsterberg, "Nothing But The Truth," 536.
47. Hugo Münsterberg, "The Third Degree," *McClure's Magazine* 29, October 1907, 614–22.
48. Münsterberg, "The Third Degree," 614.
49. Ibid., 615.
50. Ibid.
51. Ruth Benschop and Douwe Draaisma, "In Pursuit of Precision: The Calibration of Minds and Machines in Late Nineteenth-century Psychology," *Annals of Science* 57 (2000): 1–25.
52. Münsterberg, "The Third Degree," 615.
53. Ibid., 617.

54. Ken Alder, *The Lie Detectors: The History of an American Obsession* (New York: Free Press, 2007), 47.
55. Münsterberg, "The Third Degree," 619.
56. Ibid., 622.
57. "Applied Psychology and Its Possibilities," *New York Times,* September 22, 1907 V, 9.
58. Ibid.
59. Ibid.
60. Ibid.
61. White, "The Soul Machine."
62. Ibid., 12.
63. "Discovery Made by a Swiss Doctor" V, 8.
64. White, "The Soul Machine."
65. Ibid.,12.
66. "Mr. B- was asked to charge his mind with the crime of having stolen the cigar-box and the articles it contained, and to produce himself with a conscience as guilty as possible at Dr. Peterson's office the next evening, prepared to resist all efforts to extort his unholy secret from him." White, "The Soul Machine," 13.
67. O. Mezger, "Photography in the Service of the Law," *Scientific American Supplement No. 1765*, October 30, 1909, 284.
68. Ibid.
69. "Why the Great Scientist will Supersede the Great Detective," *Current Literature* 51, September 1911, 279–81.
70. "Why the Great Scientist will Supersede the Great Detective," 280.
71. "Laboratory Study of Criminals," *The Literary Digest* 45, November 16, 1912, 898.
72. Edward A. Ayres, "Measuring Thought with a Machine," *Harper's Weekly* 52, May 9, 1908, 27 (emphasis in original).
73. Ibid.
74. Ibid.
75. Ibid.
76. "Electric Machine to Tell Guilt of Criminals," *New York Times*, September 10, 1911, V, 6.
77. Ibid.
78. Ibid.
79. Ibid.
80. Raymond E. Fancher, *The Intelligence Men: Makers of the IQ Controversy* (New York: W. W. Norton and Co., 1985), 108.
81. Nicole Hahn Rafter, *Creating Born Criminals* (Champaign: University of Illinois Press, 1997), 143.
82. Ibid., 145.
83. Ibid., 136.
84. Leila Zenderland, "The Debate over Diagnosis: Henry Herbert Goddard and the Medical Acceptance of Intelligence Testing," in *Psychological Testing and American Society 1890-1930*, ed. Michael M. Sokal (New Brunswick: Rutgers University Press, 1987), 59.

85. John A. Popplestone and Marion White McPherson, "Pioneer Psychology Laboratories in Clinical Settings," in *Explorations in the History of Psychology in the United States*, ed. Josef Brozek (Lewisburg: Bucknell University Press, 1984), 241.

86. Rafter, *Creating Born Criminals*, 145.

87. Ibid., 146.

88. Ibid., 137.

89. "'There Is No Criminal Type,' Says Prison Expert." *New York Times*, November 2, 1913, SM13.

90. Arthur E. Fink, *Causes of Crime: Biological Theories in the United States* (Philadelphia: University of Philadelphia Press, 1938), 244.

91. Fink, *Causes of Crime*, 243–49; John L. Gillin, "Economic Factors in the Making of the Criminal," *Journal of Social Forces* 3, no.2 (January 1925): 248–55.

92. Fink, *Causes of Crime*, 251.

93. "Psychology Squad Latest Police Aid," *New York Times*, October 30, 1915, 5.

94. Arthur B. Reeve claimed that the addition of science to crime in detective stories "began when several writers tried to apply psychology, as developed by Prof. Hugo Muenssterberg [sic] of Harvard and Prof. Walter Dill Scott of Northwestern University, to either actual or hypothetical cases of crime." Arthur B. Reeve (1913) quoted in John Locke, ed., *From Ghouls to Gangsters: The Career of Arthur B. Reeve*, vol. 2 (Elkhorn, CA: Off-Trail Publications, 2007), 31.

95. Robert Sampson, *Yesterday's Faces: A Study of Series Characters in the Early Pulp Magazines II, Strange Days* (Bowling Green: Bowling Green University Popular Press, 1984), 4.

96. Sampson, *Yesterday's Faces*, 16.

97. Ibid., 11–12.

98. Ibid., 47. In "The Supreme Test" Wycherley subjects his subject to a blood pressure test to see if he is a true heir to a fortune or a fraudulent claimant. Sampson, *Yesterday's Faces*, 48.

99. Ibid., 16.

100. Edwin Balmer and William MacHarg, *The Achievements of Luther Trant* (Boston: Small, Maynard and Co., 1910), 4.

101. Sampson, *Yesterday's Faces*, 20.

102. Balmer and MacHarg, *The Achievements of Luther Trant*, 169.

103. Ibid., Foreword.

104. "The Red Dress," in Balmer and MacHarg, *The Achievements of Luther Trant*, 76.

105. Ibid., 88–89.

106. Ibid., 95.

107. Reeve, retitled as "The Scientific Cracksman," for *Amazing Detective Tales* (August 1930).

108. Balmer and MacHarg, *The Achievements of Luther Trant*, 355.

109. Compare, for example, "I believe that in the study of mental diseases these men are furnishing the knowledge upon which future criminologists will build to make the detection of crime an absolute certainty. Some day there will be no jury, no detectives, no witnesses, no attorneys. The state will merely submit all suspects to tests of scientific instruments like these, and as these instruments can not make mistakes or tell lies their evidence will be conclusive of guilt or innocence." "The Crimeometer" 14, in Arthur B. Reeve, *The Dream Doctor, The Craig*

Kennedy Series (New York: Heart's International Library, 1914, originally published December 1912), 217; and "There will be no jury, no horde of detectives and witnesses, no charges and countercharges, and no attorney for the defense. These impedimenta of our courts will be unnecessary. The State will merely submit all suspects in a case to the tests of scientific instruments, and as these instruments cannot be made to make mistakes nor tell lies, their evidence will be conclusive of guilt or innocence, and the court will deliver sentence accordingly." "Electric Machine to Tell Guilt of Criminals," *New York Times*, September 10, 1911 V, 6.

110. Quoted in Sampson, *Yesterday's Faces*, 23. The fictional character eerily anticipated William Moulton Marston who in real-life combined science with law and had "both the University and Third Avenue melodrama in his make-up." As Marston would later also do, Kennedy at one point gives a test to a woman while she watches a movie. See Reeve, *The Dream Doctor*.

111. Reeve was also involved with silent movie serials, some of which featured the Kennedy character. A television series, "Craig Kennedy, Criminologist," was devised in the early 1950s.

112. Sampson, *Yesterday's Faces*, 25.

113. Arthur B. Reeve, "The Truth Detector" in *The Treasure Train* (New York: Harper and Brothers, 1917), 29.

114. Arthur B. Reeve, "The Scientific Cracksman" in *The Silent Bullet* (New York: Dodd Mead, 1912).

115. Quoted in LeRoy Lad Panek, *The Origins of the American Detective Story* (Jefferson: McFarland, 2006), 105.

116. Charles Edmonds Walk, *The Yellow Circle* (New York: A. L. Burt Co., 1909), 69.

117. Arthur B. Reeve, "The Lie Detector" in *The War Terror* (Hearst's International Library Co., New York, 1915). The story was first published in *Cosmopolitan* magazine in November 1914 but was only given the title "The Lie Detector" when published as chapter 27 in Arthur B. Reeve, *The War Terror* (New York: Harper and Brothers, 1915), 273–80. It was also apparently reprinted in *The Boston Daily Globe* (February 25, 1917). Reeve also wrote stories around this time with titles like "The Detectaphone," "The Crimeometer," "The Truth Detector," and "The Love Meter." See John Locke, ed., *From Ghouls to Gangsters*, 189.

118. Cleveland Moffett, *Through the Wall* (New York: D. Appleton and Co., 1909), 171.

119. Melvin L. Severy, *The Mystery of June 13th* (New York: Dodd, Mead and Co., 1905), 523–24.

120. Although the criminologist Hans Gross had observed, "a large part of the criminalist's work is nothing more than a battle against lies," his discussion of the lie made no mention of its susceptibility to scientific detection. Particularly pertinent to criminology, however, were the lies of the "insane paralytic," the hysteric, the epileptic, and the pregnant woman. Prostitutes were apparently addicted to a particular form of lie which Lombroso and other criminologists regarded as "a professional mark of identification." The lie detector would be completely unconcerned with such "pathoformic lies" and how they signified personological types. Hans Gross, *Criminal Psychology: A Manual for Judges, Practitioners, and Students*, trans. Horace M. Kallen (Boston: Little, Brown and Co., 1911), 474–80. First published 1905.

Chapter 6. "Some of the darndest lies you ever heard": Who Invented the Lie Detector?

Epigraphs. Charles Edmonds Walk, *The Yellow Circle* (New York: A. L. Burt Co., 1909). "What electric investigative device was invented by Nova Scotia–born John Augustus Larson in 1921?" *Trivial Pursuit, Genus II* question (Canadian edition).

1. Anne Roller, "Vollmer and His College Cops," *Survey* 62 (1929): 304.
2. "Inventor of Lie Detector Traps Bride," *San Francisco Examiner*, August 9, 1922. Quoted in Ken Alder, *The Lie Detectors: The History of an American Obsession* (New York: Free Press, 2007), 11.
3. "The Lie-Detector," *The Literary Digest* 3 (December 26, 1931): 35.
4. John A. Larson, "The Lie Detector: Its History and Development," *Journal of the Michigan State Medical Society* 37 (1938): 893–97.
5. Albert A. Hopkins, "Science Trails The Criminal," *Scientific American* 146, February 6, 1932, 96.
6. "Lie Tracer is Honored," *New York Times*, January 21, 1933, 17.
7. William A. Dyche, "Science in the Detection of Crime," *The Review of Reviews*, January 1932, 52–54.
8. "Lie Detecting," *Outlook and Independent* 153, 1929, 533.
9. "Lie Detector Seals Doom of Murderer," *New York Times*, March 2, 1937, 44.
10. "Marston Advises 3 L's for Success," *New York Times*, November 11, 1937, 27.
11. William Moulton Marston, *The Lie Detector Test* (New York: Richard R. Smith, 1938), 18.
12. Robert Sampson, *Yesterday's Faces: A Study of Series Characters in the Early Pulp Magazines. Vol. 2, Strange Days* (Bowling Green: Bowling Green University Popular Press, 1984), 27.
13. "Science Devises a Painless "3rd Degree,'" *Current Opinion* 76, April 1924, 474.
14. Frederick L. Collins, "The Future Looks Dark for Liars," *Collier's*, August 16, 1924, 7, 26.
15. Ibid., 7.
16. Gene E. Carte and Elaine H. Carte, *Police Reform in the United States: The Era of August Vollmer* (Berkeley: University of California Press, 1975), 2.
17. Marston, *The Lie Detector Test*.
18. "Marston, William Moulton," *Encyclopedia of American Biography*, n.s. 7 (New York: The American Historical Society, 1937), 23.
19. Marvin S. Bowman, "New Machine Detects Liars," *Boston Sunday Advertiser*, May 8, 1921, B3.
20. Ibid.
21. "William Moulton Marston," *Harvard Class of 1915 25th Annual Report* (Pusey Library, Harvard University Archives, 1940), 480–81.
22. William Moulton Marston, "Have a Vacation Every Day," *The Rotarian* 56, January 1940, 26.
23. "Marston, William Moulton," *The National Cyclopaedia of American Biography, Current Volume, E, 1937–38* (New York: James T. White and Co., 1938), 29.

24. William Moulton Marston, "Lie Detection: Its Bodily Basis and Test Procedure," in *Encyclopedia of Psychology*, ed. Philip Lawrence Harriman (New York: Philosophical Library, 1946), 358.

25. Eugene B. Block. *Lie Detectors: Their History and Use* (New York: David McKay Co., 1977), chap 4.

26. Eloise Keeler, *The Lie Detector Man: The Career and Cases of Leonarde Keeler* (Boston: Telshare Publishing, 1984), 2.

27. Fred E. Inbau, "Scientific Crime Detection: Early Efforts in Chicago," an oral history conducted in 1972 by Gene Carte, in *August Vollmer: Pioneer in Police Professionalism, Vol. 2*. (Regional Oral History Office, The Bancroft Library, University of California, Berkeley, 1983), 6.

28. William W. Turner, *Invisible Witness: The Use and Abuse of the New Technology of Crime Investigation* (New York: The Bobbs-Merrill Co., 1968), 33.

29. Dwight G. McCarty, "Detecting the Liar," chap. 12 in *Psychology and the Law* (Englewood Cliffs: Prentice-Hall, Inc., 1960).

30. Thomas J. Deakin, *Police Professionalism: The Renaissance of American Law Enforcement* (Springfield: Charles C. Thomas, 1988), 98.

31. "Marston, William Moulton (1893–1947)," in *The World Encyclopedia of Comics*, ed. Maurice Horn, 2 vols. (New York: Chelsea House Publishers), 480–81.

32. Trina Robbins, *The Great Women Superheroes* (Northampton, MA: Kitchen Sink Press, 1996), 4.

33. Jim Korkis, "William Moulton Marston," *Comic Book Marketplace* 23 (1995): 46.

34. Les Daniels, *Wonder Woman: The Complete History* (London: Titan Books, 2000). Echoes of Marston's claim to priority are evident in his son's recent comments: "I know there was some controversy as to whether he was the first to discover the relationship [between blood pressure and lying], but he did much basic research and had developed a crude working apparatus while still at Harvard" (12).

35. Gerard Jones, *Men of Tomorrow: Geeks, Gangsters and the Birth of the Comic Book* (London: William Heinemann, 2005), 206.

36. See, for example, Vollmer's proposed syllabus for the Berkeley School of Police: A. Vollmer and A. Schneider, "The School for Police as Planned at Berkeley," *Journal of the American Institute of Criminal Law and Criminology* 7, no. 6 (1917): 877–98.

37. Gene Carte, "Introduction," in *August Vollmer: Pioneer in Police Professionals, Vol. 2*, viii.

38. William Moulton Marston, "Systolic Blood Pressure Symptoms of Deception," *Journal of Experimental Psychology* 2 (1917): 117–63; William Moulton Marston, "Reaction Time Symptoms of Deception," *Journal of Experimental Psychology* 3 (1920): 72–87; and William Moulton Marston, "Psychological Possibilities in the Deception Tests," *Journal of Criminal Law and Criminology* 11 (1921): 551–70.

39. Marston, "Systolic Blood Pressure Symptoms," 162.

40. Ibid., 163.

41. Marston, "Psychological Possibilities in the Deception Tests."

42. Bowman, "New Machine Detects Liars," B3.

43. Ibid.

44. John A. Larson, *Lying and its Detection: A Study of Deception and Deception Tests* (Glen Ridge: Patterson Smith, 1932), 261–62 (emphasis added).

45. In 1938, Marston claimed that he had initially employed a continuous measure of blood pressure, but he later "gave up the continuous record because I believed constant pressure on the subject's arm altered his blood pressure." Marston, *The Lie Detector Test*, 98.

46. John A. Larson, "The Lie Detector: Its History and Development," 894.

47. John A. Larson, "The Cardio-Pneumo-Psychogram and Its Use in the Study of the Emotions, with Practical Application," *Journal of Experimental Psychology* 5 (1922): 323–28. Keeler would later describe his polygraph as a "pneumo-cardio-sphygmo-galvanograph." Leonarde Keeler, "Debunking the 'Lie-Detector,'" *Journal of Criminal Law, Criminology and Police Science* 25 (1934–35): 157.

48. John A. Larson, "Introduction," in Marston, *The Lie Detector Test*.

49. Ibid.

50. William Moulton Marston, "Reaction Time Symptoms of Deception," *Journal of Experimental Psychology* 3 (1920): 72–87.

51. Bowman, "New Machine Detects Liars," B3.

52. Marston, "Reaction Time Symptoms of Deception," 87.

53. Ibid., 83.

54. Collins, "The Future Looks Dark for Liars."

55. Tom White, "Every Crime Is Entrenched Behind a Lie," *Scientific American* 133, 1925, 298–99.

56. "Transmission of Criminal Traits," *The Green Bag* 3, 1891, 215–16; Edmund R. Spearman, "Criminals and Their Detection," *The New Review* 9, 1893, 65–84; Sanger Brown, "Responsibility in Crime from the Medical Standpoint," *The Popular Science Monthly* 46, 1894–95, 154–64.

57. "Inventor of Lie Detector Traps Bride," quoted in Ken Alder, *The Lie Detectors*, 11.

58. "Machine Tests Veracity," *New York Times*, June 11, 1922, 5.

59. "Science Devises a Painless '3rd Degree'"; Collins, "The Future Looks Dark for Liars."

60. Marston, *Lie Detector Test*, 24; Larson, "The Cardio-Pneumo-Psychogram and Its Use in the Study of the Emotions."

61. "New Lie Detector was Used on Green," *New York Times*, January 15, 1937, 3.

62. "Lie Detection: Device Invented by Priest Wins First Court Recognition," *Newsweek* 11, April 11, 1938, 26.

63. Walter G. Summers, "Science Can Get the Confession," *Fordham Law Review* 8 (1939): 334–54.

64. "New Lie Detector Was Used on Green." Summers' machine was depicted in an English weekly magazine serial in the late 1930s in answer to the question "What Is a Lie Detector?" in *Everybody's Enquire Within*, ed. Charles Ray, vol. 1 (London: The Amalgamated Press, Ltd., n.d. [probably 1939]).

65. Although it has become an icon of American popular culture, at least one Russian psychologist has also claimed to have invented the lie detector. In his reminiscences about Soviet psychology, A. R. Luria proclaimed that his early work "turned out to be of practical value to criminologists, providing them with an early model of a lie detector. Aleksandr R.

Luria, *The Making of Mind: A Personal Account of Soviet Psychology* (Cambridge, MA: Harvard University Press, 1979), 35–36.

66. "A Machine to Measure Lies," *Look,* January 4, 1938, 29.

67. Ibid.

68. Larson, "Introduction," in Marston, *The Lie Detector Test.*

69. Keeler, *The Lie Detector Man,* 2.

70. Marston, *The Lie Detector Test,* 28.

71. Ibid., 24–25.

72. Ibid., 25–26.

73. Sean Dennis Cashman, *America in the Age of the Titans: The Progressive Era and World War I* (New York: New York University Press, 1988), 10–18.

74. On the emergence of the notion of the heroic inventor in Britain, see Christine MacLeod, *Heroes of Invention: Technology, Liberalism and British Identity, 1750–1914* (Cambridge: Cambridge University Press, 2007).

75. Ibid., 20.

76. Arthur S. Link, *American Epoch: A History of the United States since the 1890s* (New York: Alfred A. Knopf, 1955), 305–6.

77. Thomas P. Hughes, *American Genesis: A Century of Invention and Technological Enthusiasm, 1870–1970* (Chicago: University of Chicago Press, 2004), 3.

78. Cashman, *America in the Age of the Titans,* 26.

79. Keeler filed his patent on July 30, 1925. His invention permitted a simultaneous recording of both the systolic/diastolic cardiac cycle as well as slower, irregular oscillations "which may be superimposed on a considerable number of cardiac cycles." See "United States Patent Office: Leonarde Keeler of Berkeley, California. Apparatus for Recording Arterial Blood Pressure," *Polygraph: Journal of the American Polygraph Association* 3 (1974): 210–15.

80. On scientific personae, see Lorraine Daston and H. Otto Sibum, "Introduction: Scientific Personae and Their Histories," *Science in Context* 16, no. 1 (2003): 1–8.

81. George Basalla, *The Evolution of Technology* (Cambridge: Cambridge University Press, 1988), 21–28.

82. Basalla, *The Evolution of Technology,* 57. See also Bruno Latour, *Science in Action: How to Follow Scientists and Engineers through Society* (Cambridge, MA: Harvard University Press, 1987), especially chap. 3.

83. The basis of "Keeler's mystique," to use Ken Alder's apt term, was his patent. But even here the mystique was unfounded because not only did the patent involve merely a minor detail concerning tambour design, but the innovation, Alder suggests, could in fact be credited to the work of Keeler's former colleagues Hiram Edwards and Charles Sloan (Alder, *The Lie Detectors,* 78). A further irony was that Keeler's metal tambours (unlike C. D. Lee's all-rubber ones) "were prone to fracture and required frequent calibration" (ibid., 123).

84. It also accounts for the acrimonious relationships between the original pioneers. This point is explored at greater length in chap. 6.

85. Inbau, "Scientific Crime Detection," 6.

86. Although Larson does say that "Keeler's polygraph has many mechanical imperfections, including those of the driving mechanism." Larson, *Lying and Its Detection,* 279.

87. Leonarde Keeler to August Vollmer, September 19, 1932, August Vollmer Papers, ca. 1918–1955, Bancroft Library, University of California, Berkeley, California (hereafter AVP).
88. Leonarde Keeler to August Vollmer, March 19, 1934, AVP.
89. Marston, *The Lie Detector Test*, 27, 50.
90. Ibid., 108.
91. Ibid., 145.
92. Ibid., 78.
93. Leonarde Keeler to August Vollmer, March 28, 1938, AVP.
94. Fred E. Inbau, "*The Lie Detector Test* by William Moulton Marston," *Journal of Criminal Law, Criminology and Police Science* 29 (1938): 305.
95. Ibid.
96. Ibid., 307–8.
97. John A. Larson, "*Lie Detection and Criminal Investigation* by Fred E. Inbau," *Fordham Law Review* 12 (1943): 307–10.
98. Ibid., 308.
99. Inbau, "Scientific Crime Detection," 8.
100. Ibid., 5.
101. Ibid., 8.
102. Ibid., 5.
103. Larson to Vollmer, June 2, 1951, AVP.
104. Ibid. (emphasis in original).
105. F. Ellenberger, *The Discovery of the Unconscious: The History and Evolution of Dynamic Psychiatry* (New York: Basic Books, 1970), 732.
106. Mathew Hale, *Human Science and Social Order: Hugo Münsterberg and the Origins of Applied Psychology* (Philadelphia: Temple University Press, 1980), 120.
107. Jutta Spillmann and Lothar Spillmann, "The Rise and Fall of Hugo Münsterberg," *Journal of the History of the Behavioral Sciences* 29 (1993): 329.
108. Larson to Vollmer, June 2, 1951, 3, AVP.
109. Larson, *Lying and its Detection*, 190.
110. Bruno Latour has argued that one of technoscience's most powerful features is its portability. Bruno Latour (1990), "Drawing Things Together," in *Representation in Scientific Practice*, ed. Michael Lynch and Steve Woolgar (Cambridge, MA: MIT Press, 1990), 19–68.
111. Alder on Keeler's software.
112. Latour, *Science in Action*, 81–82.
113. To this extent, then, the fiction of invention may be regarded as an "origin myth." As a number of historians have argued, origin myths are far from unusual in historical writing in general and in the history of psychology in particular. On the notion of "invented tradition" see Eric J. Hobsbawm and Terence O. Ranger, eds., *The Invention of Tradition* (Cambridge: Cambridge University Press, 1983); Ben Harris, "Whatever Happened to Little Albert?" in Ludy Benjamin, ed., *A History of Psychology: Original Sources and Contemporary Research* (New York: McGraw-Hill, 1988), 424–31.

Chapter 7. "A trick of burlesque employed . . . against dishonesty": The Quest for Euphoric Security

Epigraph. Roland Barthes, "Einstein's Brain," in *Mythologies* (London: Paladin Books, 1985), 77. Max Weber, "The Sociology of Charismatic Authority/The Nature of Charismatic Authority and Its Routinization," in P. David Marshall, ed., *The Celebrity Culture Reader* (London: Routledge, 2006), 59.

1. Fred E. Inbau, "The 'Lie-Detector,'" *Scientific Monthly* 40, 1935, 81–87.
2. Paul V. Trovillo, "A History of Lie Detection," *Journal of Criminal Law and Criminology* 30 (1939): 848.
3. Ibid., 858.
4. William Moulton Marston, *The Lie Detector Test* (New York: Richard R. Smith, 1938), 7.
5. Nealis O'Leary, "A Criminologist to the Rescue," *The Literary Digest* 118, October 6, 1934, 22.
6. "Science Devises a Painless '3rd Degree,'" *Current Opinion* 76, April 1924, 474.
7. "The Third Degree," *Look,* August 31, 1937, 16–17; "How a Lie Detector Works," *Look,* August 31, 1937, 23.
8. Henry Morton Robinson, "Science Gets the Confession," *Forum and Century* 93, 1935, 15.
9. Geoffrey C. Bunn, "Spectacular Science: The Lie Detector's Ambivalent Powers," *History of Psychology* 10, no. 2 (2007): 160.
10. Z. Chafee, W. H. Pollack, and C. S. Stern, *The Third Degree* (New York: Arno Press, 1969). Reprint of *United States Wickersham Commission Report No.11: Report on Lawlessness in Law Enforcement,* 1931.
11. E. J. Hopkins, "The Lawless Arm of the Law," *The Atlantic Monthly* 148, 1931, 284–85.
12. See, for example, "Examination by Torture," *The Outlook,* May 30, 1908, 237–38; Richard Sylvester, "The Treatment of the Accused," *Annals of the American Academy of Political and Social Science* 36 (1910): 16–19.
13. Hugo Münsterberg, *On the Witness Stand: Essays on Psychology and Crime* (New York: Clark Boardman, 1927), 74. First published in 1908. Münsterberg's denunciations of the third degree inspired the humanitarian Charles Klein to write a play about the subject. Produced in New York in 1909, it was described as being "one of the cleverest and most startling experiments in melodramatics." "The Third Degree," *Current Literature* 47, October 1909, 427.
14. Marston, *The Lie Detector Test,* 97.
15. William A. Dyche, "Science in the Detection of Crime," *The Review of Reviews,* January 1932, 52.
16. Thomas H. Jaycox, "Scientific Detection of Lies," *Scientific American* 156, June 1937, 371.
17. Alva Johnston, "The Magic Lie Detector I," *Saturday Evening Post* 216, April 15, 1944, 9.
18. Alva Johnston, "The Magic Lie Detector III," *Saturday Evening Post* 216, April 29, 1944, 102.
19. Kenneth Murray, "Two Simple Ways to Make a Lie Detector," *Popular Science Monthly* 128, 1936, 63, 98–99.
20. Dyche, "Science in the Detection of Crime," 52–53.

21. Inbau, "The 'Lie-Detector,'" 83.
22. "Lie Detection," *Living Age* 348, March 1935, 92.
23. Robinson, "Science Gets the Confession," 16.
24. Henry F. Pringle, "How 'Good' is Any Lie?," *Reader's Digest* (American Edition) 29, November 1936, 76.
25. "Lie Detector: Marks in Ink Final Judges for Murder Case," *Newsweek* 9, March 13, 1937, 34.
26. "Lie Detector Casts Doubt on Constable," *New York Times*, November 18, 1937, 19.
27. "Wichita's Use of the Lie Detector," *The American City* 51, December 1936, 91.
28. Thomas H. Jaycox, "Lies-Truths," *Scientific American* 161, July 1939, 8.
29. J. P. McEvoy, "The Lie Detector Goes into Business," *Reader's Digest* (American Edition) 38, February 1941, 71.
30. Johnston, "The Magic Lie Detector I," 73.
31. William Moulton Marston, "Systolic Blood Pressure Symptoms of Deception," *Journal of Experimental Psychology* 2 (1917): 162.
32. William Moulton Marston, "Psychological Possibilities in the Deception Tests," *Journal of Criminal Law and Criminology* 11 (1921): 568.
33. Harold E. Burtt, "The Inspiration-Expiration Ratio During Truth and Falsehood," *Journal of Experimental Psychology* 4 (1921): 23.
34. Leonarde Keeler, "A Method For Detecting Deception," *American Journal of Police Science* 1 (1930): 44.
35. Fred E. Inbau, "Scientific Evidence in Criminal Cases II: Methods of Detecting Deception," *Journal of Criminal Law, Criminology and Police Science* 24 (1933–34): 1147.
36. Trovillo, "A History of Lie Detection," 878.
37. Walter G. Summers, "Science Can Get the Confession," *Fordham Law Review* 8 (1939): 338–40.
38. John Larson to August Vollmer, January 22, 1931, August Vollmer Papers, ca. 1918–1955, Bancroft Library, University of California, Berkeley, California (hereafter AVP).
39. Johnston, "The Magic Lie Detector I," 73.
40. Pringle, "How 'Good' Is Any Lie?," 75–76.
41. Leonarde Keeler to August Vollmer, July 7, 1948, AVP.
43. "Principal Wrecks His 'Lie Detector,'" *New York Times*, September 25, 1936, 25.
44. Jaycox, "Scientific Detection of Lies," 370.
45. Murray, "Two Simple Ways to Make a Lie Detector," 63.
46. "Crime Wave? Try Vollmer's Educated Cops," *The Literary Digest* 101, June 29, 1929, 33–37.
47. "'Lie Detector' Doesn't," *Science News Letter* 49, March 30, 1946, 207.
48. "'Lie Detector' Traps Philadelphia Youth," *New York Times*, July 7, 1931, 23.
49. Johnston, "The Magic Lie Detector I," 9, 73.
50. Edwin Balmer and William MacHarg, *The Achievements of Luther Trant* (Boston: Small, Maynard & Co., 1910), 169.
51. Dyche, "Science in the Detection of Crime," 53.
52. "Detecting Liars," *The Literary Digest* 114, October 22, 1932, 22.

53. "Catching Criminals With 'Lie-Detector,'" *The Literary Digest* 119, February 23, 1935, 17.
54. Inbau, "The 'Lie-Detector,'" 83.
55. "Polygraph Proof," 10.
56. Pringle, "How 'Good' Is Any Lie?," 75.
57. "Lie Detector Seals Doom of Murderer," *New York Times*, March 2, 1937, 44.
58. A. A. Lewis, "Looking for an Honest Man," *Scientific Monthly* 49, 1939, 269, 270.
59. Jaycox, "Scientific Detection of Lies," 372.
60. Robinson, "Science Gets The Confession," 17.
61. "'Lie Detector' Tried on Bomb Suspects," *New York Times*, July 25, 1931, 15.
62. "Polygraph Proof," 10.
63. "Lie Detector: Marks in Ink Final Judges for Murder Case," 34.
64. Ken Alder, *The Lie Detectors: The History of an American Obsession* (New York: Free Press, 2007), 82.
65. Johnston, "The Magic Lie Detector I," 9.
66. Rufus P. Turner, "Crime Detection," *Radio News* 30, 1943, 20.
67. McEvoy, "The Lie Detector Goes into Business," 70.
68. Pringle, "How 'Good' Is Any Lie?," 75, 76.
69. Jaycox, "Lies-Truths," 9.
70. Lewis, "Looking for an Honest Man," 268.
71. Pringle, "How 'Good' Is Any Lie?," 76.
72. "Spurn Lie Detector, Detectives Demoted," *New York Times*, July 13, 1938, 4.
73. Johnston, "The Magic Lie Detector III," 101.
74. "'Lie Detector' Used in Maine," *New York Times*, October 12, 1935, 3.
75. "Lie Detector Proves Bloodhounds Are Liars," *New York Times*, November 11, 935, 6.
76. Eloise Keeler, *The Lie Detector Man: The Career and Cases of Leonarde Keeler* (Boston: Telshare Publishing, 1984), chap. 13.
77. "Lie-Detector Test Asked By Prisoner," *New York Times*, January 6, 1935, Section IV, 4.
78. "Hauptmann Pleads for a Truth Test," *New York Times*, December 17, 1935, 3.
79. "Gov. Hoffman Urges Lie-Detector Test," *New York Times*, January 24, 1936, 40. According to Ken Alder, Orlando Scott, John Larson, and Keeler were all angling to submit Hauptmann to a lie detector test. Alder, *The Lie Detectors*, 148.
80. "Coogan and Fiancée Robbed in Chicago," *New York Times*, February 13, 1936, 25.
81. "Londes Bans Lie Test," *New York Times*, November 8, 1949, 38.
82. See, for example, Inbau, "The 'Lie-Detector.'"
83. Marston, *The Lie Detector Test*, 177.
84. Johnston, "The Magic Lie Detector III," 102.
85. "Pros and Cons of the Lie Detector," *New York Times*, August 22, 1948, IV, 7.
86. Dyche, "Science in the Detection of Crime," 53.
87. "Admits Lie Detector in Compensation Case," *New York Times*, August 20, 1935, 5.
88. "Lie Detector Seals Doom of Murderer," 44.
89. "Polygraph Proof," 9.
90. "Lie Detector Gains in Use," *New York Times*, July 28, 1935, IV, 11.

91. "Lie Detector Proves Bloodhounds Are Liars," 6.
92. "Truth Wanted," *Time* 43, January 10, 1944, 60.
93. "Lie-Detecting," *Outlook and Independent* 153, December 4, 1929, 533.
94. Jaycox, "Scientific Detection of Lies," 373.
95. "In the Driftway," *The Nation* 129, December 11, 1929, 719.
96. "Lie Detection," *Living Age* 348, March 1935, 92.
97. Tom White, "Every Crime Is Entrenched Behind a Lie," *Scientific American* 133, 1925, 298–99.
98. Marston, *The Lie Detector Test*, "Acknowledgments."
99. Balmer and MacHarg, *The Achievements of Luther Trant*, 25.
100. Trovillo, "A History of Lie Detection," 876.
101. Marvin S. Bowman, "New Machine Detects Liars," *Boston Sunday Advertiser*, May 8, 1921, B3.
102. Frederick L. Collins, "The Future Looks Dark for Liars," *Collier's*, August 16, 1924, 7.
103. Anne Roller, "Vollmer and His College Cops," *Survey* 62, 1929, 304. The photograph was a still from an educational movie made in Berkeley in which Chief Vollmer and his staff played star roles.
104. John A. Larson, *Lying and its Detection: A Study of Deception and Deception Tests* (Glen Ridge: Patterson Smith, 1932), fig. 1.
105. O'Leary, "A Criminologist to the Rescue," 22.
106. Fred E Inbau, "The 'Lie-Detector,'" *Scientific Monthly* 40, 1935, 82; Albert A. Hopkins, "Science Trails The Criminal," *Scientific American* 146, February 6, 1932, 96.
107. "The Week in Science: Detecting Lies of Criminals," *New York Times*, January 13, 1935, Section VIII, 4.
108. Murray, "Two Simple Ways To Make A Lie Detector," 63.
109. Wayne Biddle, "The Deception of Detection," *Discover* 7, March 1986, 26.
110. Johnston, "The Magic Lie Detector I," 9.
111. "Lie Detector Tests on Workers," *Business Week*, April 28, 1951, 24.
112. Geoffrey C. Bunn, "The Hazards of the Will to Truth: A History of the Lie Detector," PhD diss., York University, Toronto, 1998, 245.
113. John E. Reid and Fred E. Inbau, *Truth and Deception: The Polygraph ("Lie Detection") Technique* (Baltimore: Williams and Wilkins, 1977), 5.
114. James Allan Matté, *The Art and Science of the Polygraph Technique* (Springfield, IL: Charles C. Thomas, 1980), fig. 67, 132; Paul L. Wilhelm and F. Donald Burns, *Lie Detection with Electrodermal Response*, 5th ed. (Michigan City, IN: B and W Associates, 1954).
115. Trovillo, Keeler and Reid, and Inbau all employed office secretaries for the posed photographs. See, for example, Keeler, *Lie Detector Man*, 69.
116. Marston, *The Lie Detector Test*, 113, 115, 114.
117. "A Machine to Measure Lies," *Look*, January 4, 1938, 29.
118. "It Really Understands Women," newspaper clipping (May 1938), in Boder Museum Papers, Archives of the History of American Psychology, University of Akron, OH.
119. "The metaphorical associations of (un)veiling are rich and diverse, going far beyond their direct connections with scientific knowledge, encompassing religion (nuns, ideas of revelation, the cover of a chalice), clothing, crime, mystery, horror and deceit of all kinds."

Ludmilla Jordanova, "Nature Unveiling before Science," in *Sexual Visions: Images of Gender in Science and Medicine between the Eighteenth and Twentieth Centuries* (New York: Harvester Wheatsheaf, 1989), 91. See also Carolyn Merchant, *The Death of Nature: Women, Ecology, and the Scientific Revolution* (San Francisco: Harper and Row, 1983).

120. See for example Matté, *The Art and Science of the Polygraph Test*, fig. 67, 132; and Reid and Inbau, *Truth and Deception*, fig. 1, 5. Reid and Inbau also discuss some of the potential hazards of attaching the pneumographic tube to women (22, footnote 28).

121. Reid and Inbau, *Truth and Deception*, 152.

122. Matté, *The Art and Science of the Polygraph Technique*, 133; Reid and Inbau, *Truth and Deception*, 7.

123. "Popular electronics" for example, traditionally a male activity, has often provided instructions on how to build a lie detector. See Murray, "Two Simple Ways to Make a Lie Detector"; Edwin Bohr, "Lie Detector," *Radio and Television News* 49, 1953, 56–57.

124. Dori J. Pearl, "Why a Female Polygraphist?" *The Journal of Polygraph Science* 10 (September-October 1975): 1.

125. Roland Barthes, "Einstein's Brain," in *Mythologies* (London: Paladin Books, 1985), 77.

Chapter 8. "A bally hoo side show at the fair": The Spectacular Power of Expertise

Epigraphs. "Lie Detector 'Tells All,'" *Life*, November 21, 1938, 65. Eloise Keeler, *The Lie Detector Man: The Career and Cases of Leonarde Keeler* (Boston: Telshare Publishing, 1984), 28.

1. "How a Lie Detector Works," *Look*, August 31, 1937, 23.

2. William Moulton Marston, "Would YOU Dare Take These Tests?" *Look*, December 6, 1938, 16.

3. Frederick L. Collins, "The Future Looks Dark for Liars," *Collier's*, August 16, 1924, 26.

4. On Marston see Geoffrey C. Bunn, "The Lie Detector, *Wonder Woman* and Liberty: The Life and Work of William Moulton Marston," *History of the Human Sciences* 10 (1997): 91–119.

5. The rejection of the systolic blood pressure deception test in the Frye v. United States case on the grounds of its insufficient acceptance among scientific authorities was a landmark ruling, because it inadvertently established the legal criteria for the admissibility of scientific evidence. In 1979 the Kansas Supreme Court asserted: "The Frye test has been accepted as the standard in practically all of the courts of this country which have considered the question of the admissibility of new scientific evidence." J. E. Starrs, "'A Still-Life Watercolor': Frye v. United States," *Journal of Forensic Sciences* 27 (1982): 685.

6. "William Moulton Marston," *Harvard Class of 1915, 25th Anniversary Report* (Cambridge, MA: Harvard University Archives, Pusey Library, 1940), 480–82.

7. William Moulton Marston, "The Psychonic Theory of Consciousness," *Journal of Abnormal and Social Psychology* 21 (1926): 161–69; Marston, "Consciousness, Motation, and Emotion," *Psyche* 29 (1927): 40–52; Marston, "Motor Consciousness as a Basis for Emotion," *Journal of Abnormal and Social Psychology* 22 (1927): 140–50; Marston, "Primary Colours and Primary Emotions," *Psyche* 30 (1927): 4–33; Marston, "Primary Emotions," *Psychological Review* 34 (1927): 336–63; Marston, "Materialism, Vitalism and Psychology," *Psyche* 31 (1928): 15–34.

8. "Blondes Lose Out in Film Love Test," *New York Times,* January 31, 1928, 25.

9. "Marston, William Moulton," *Encyclopedia of American Biography,* n.s. 7 (New York: The American Historical Society, 1937), 24.

10. William Moulton Marston, "Science Derides the 'Love-Slave' Verdict, Crying '*Woman* is the *Man's* Love-Master,'" *Chicago American,* December 22, 1934, magazine section.

11. Ibid. (emphasis in original).

12. Jutta Spillmann and Lothar Spillmann, "The Rise and Fall of Hugo Münsterberg," *Journal of the History of the Behavioral Sciences* 29 (1993): 329.

13. William Moulton Marston, *The Lie Detector Test* (New York: Richard R. Smith, 1938), 46–47.

14. William Moulton Marston, "Why 100,000,000 Americans Read Comics," *American Scholar* 13 (1944): 42–43.

15. Mike Benton, *The Comic Book in America: An Illustrated History* (Dallas: Taylor Publishing Co., 1989), 32.

16. Wonder Woman appeared in *All Star Comics, Sensation Comics, Comic Calvacade,* and *Wonder Woman.* See Benton, *The Comic Book in America,* 32–36.

17. Peggy le Boutillier, "The Amazons are Coming," *The Woman,* July 1943, 66–67. In 1944 the character was syndicated: "Wonder Woman Joins Superman-Batman in National Newspaper Syndication," *Independent News* 1, no. 3 (April 1944): 3, MSS 1618B RB NMAH William Moulton Marston, "'Wonder Woman': Selected Continuities" (Washington, DC: Smithsonian Institution Libraries, 1942–1970).

18. For a selection of Marston's stories see Charles Moulton [William Moulton Marston] *Wonder Woman* (New York: Holt, Rinehart and Winston and Warner Books, 1972. First published 1943–49).

19. Bunn, "The Lie Detector, *Wonder Woman* and Liberty."

20. His sexual "endocrine" theories were in keeping with what most popular psychology was promoting during the 1920s and 1930s. See John C. Burnham, "The New Psychology: From Narcissism to Social Control," in *Paths into American Culture: Psychology, Medicine, and Morals* (Philadelphia: Temple University Press, 1988. First published 1968); Burnham, *How Superstition Won and Science Lost: Popularizing Science and Health in the United States* (New Brunswick: Rutgers University Press, 1987), 98–99.

21. "Lie Detector 'Tells All,'" 65.

22. Marston might have persuaded the readers of *Life* magazine of the lie detector's capabilities, but he didn't convince the Federal Bureau of Investigation. Writing to explain how the Gillette advert had come about, an FBI Special Agent told FBI Director J. Edgar Hoover in July 1939 that Marston "stood to make around thirty thousand dollars for his part in the entire scheme." At the bottom of the letter someone, possibly Hoover, scrawled "I always thought this fellow Marston was a phony and this proves it." FBI File on William Moulton Marston, http://antipolygraph.org/documents/marston-fbi-file.pdf (accessed April 3, 2008).

23. For similar "confessional" autobiographical works by polygraphists see Robert J. Ferguson Jr., *The Scientific Informer* (Springfield, IL: Charles C. Thomas, 1971); and Chris Gugas, *The Silent Witness: A Polygraphist's Casebook* (Englewood Cliffs: Prentice-Hall, Inc., 1979).

24. Marston, *The Lie Detector Test,* 87.

25. Ibid., 7.

26. Ibid., chap. 2.

27. Historians doubt the existence of a "crime wave" in 1930s, attributing the sense of national emergency to a few spectacular and well-publicized events like the St. Valentine's Day massacre and the Lindbergh baby kidnapping. Supported by the press, Hoover's FBI did much to whip up national hysteria. In 1933 two criminologists asserted, "No support is found for the belief that an immense crime wave has engulfed the United States." Quoted in Samuel Walker, *A Critical History of Police Reform: The Emergence of Professionalism* (Lexington, MA: Lexington Books, 1977), 152.

28. Marston, *The Lie Detector Test*, 15.

29. Ibid., 17, 16, 99, 103, 29, 132, 133, 142.

30. Charles Keeler to August Vollmer, March 14, 1930, August Vollmer Papers, ca. 1918–1955, Bancroft Library, University of California, Berkeley, California (hereafter AVP).

31. Eloise Keeler, *The Lie Detector Man: The Career and Cases of Leonarde Keeler* (Boston: Telshare Publishing, 1984), 12.

32. Ken Alder, *The Lie Detectors: The History of an American Obsession* (New York: Free Press, 2007), 55.

33. Joseph G. Woods, "Introduction," in *Law Enforcement in Los Angeles: Los Angeles Police Department Annual Report, 1924*, ed. August Vollmer (New York: Arno Press, 1974.)

34. It was Jacques Loeb who had introduced Vollmer to European criminologists such as Hans Gross. See Gene E. Carte and Elaine H. Carte, *Police Reform in the United States: The Era of August Vollmer* (Berkeley: University of California Press, 1975). Alfred Parker suggests that it was Gross' discussion of lying that inspired Vollmer to instruct his department to build a lie detector. Alfred Parker, *The Berkeley Police Story* (Springfield: Charles C. Thomas, 1972).

35. Walker, *A Critical History of Police Reform*, chap. 4.

36. Carte and Carte, *Police Reform*, 45.

37. Thomas J. Deakin, *Police Professionalism: The Renaissance of American Law Enforcement* (Springfield: Charles C. Thomas, 1988), 89.

38. Walker, *A Critical History of Police Reform*, 82.

39. Carte and Carte, *Police Reform*, 57.

40. Quoted in Woods, *Law Enforcement in Los Angeles*, 162.

41. August Vollmer, *The Police and Modern Society* (Montclair: Patterson Smith, 1936), 222.

42. James F. Richardson, *Urban Police in the United States* (Port Washington, NY: National University Publications, 1974), 135.

43. Walker, *A Critical History of Police Reform*, chap. 4.

44. Alder, *The Lie Detectors*, 76–77.

45. Ibid., 79.

46. Charles Keeler to August Vollmer, December 23, 1929, AVP.

47. Leonarde Keeler to August Vollmer, September 17, 1937, AVP.

48. Leonarde Keeler to August Vollmer, March 28, 1938, AVP.

49. Alder, *The Lie Detectors*, 239–40.

50. Ibid., 68.

51. Collins, "The Future Looks Dark for Liars," 7.

52. Ibid., 26.

53. Leonarde Keeler, "A Method For Detecting Deception," *American Journal of Police Science* 1 (1930).
54. Charles Keeler to August Vollmer, November 26, 1929, AVP.
55. Keeler, *The Lie Detector Man*, 28.
56. "Guilty By Lie Detector," *New York Times*, May 24, 1930, 2.
57. Leonarde Keeler, "'The Canary Murder Case': The Use of the Deception Test to Determine Guilt," *American Journal of Police Science* 1 (1930): 381–86.
58. "Lie Detector 'Clears' Winkler," *New York Times*, November 26, 1931, 56.
59. Keeler, *The Lie Detector Man*, chap. 13.
60. "Lie Tracer is Honored," *New York Times*, January 21, 1933, 17.
61. Keeler, *The Lie Detector Man*, 81.
62. Ibid., 83.
63. Fred E. Inbau, "The 'Lie-Detector,'" *Scientific Monthly* 40, 1935, 81–82.
64. Keeler, *The Lie Detector Man*, 162.
65. Quoted in Keeler, *The Lie Detector Man*, 160–61.
66. "Paramount Pictures Presents Popular Science [1944]," in *The Smithsonian Institution Presents: Invention*, a production of The Discovery Channel and Koch TV Productions, Inc., in association with Beyond Productions PTY Ltd.
67. The New York Police Department was inspired to invest $7000 in a similar "Mobile Laboratory Truck." See "Clue Wagon," *New Yorker* 19, December 4, 1943, 28.
68. William A. Dyche, "Science in the Detection of Crime," *The Review of Reviews*, January 1932, 52.
69. Henry Morton Robinson, "Science Gets the Confession," *Forum and Century* 93, 1935, 15.
70. Keeler, *The Lie Detector Man*, 35.
71. Ibid., 97–98.
72. Fred Inbau later recalled that the Chicago World's Fair crime detection exhibit was "a good ad for the university and for the whole cause of scientific crime detection." Fred E. Inbau, "Scientific Crime Detection: Early Efforts in Chicago," in Gene Carte, *August Vollmer: Pioneer in Police Professionalism* 2 (Berkeley: University of California Bancroft Library, 1983), 4.
73. Nealis O'Leary, "A Criminologist to the Rescue," *The Literary Digest* 118, October 6, 1934, 22.
74. "Lie Detection," *Living Age* 348, March 1935, 92.
75. "New Crime-Detection Laboratories," *The American City* 49, October 1934, 13.
76. Deakin, *Police Professionalism*, 156–57.
77. Leonarde Keeler to August Vollmer, July 19, 1932, AVP.
78. Leonarde Keeler to August Vollmer, February 12, 1935, AVP.
79. Leonarde Keeler to August Vollmer, March 19, 1934, AVP.
80. Alder, *The Lie Detectors*, 140.
81. Leonarde Keeler to August Vollmer, March 19, 1934, AVP.
82. Alder, *The Lie Detectors*, 138.
83. Leonarde Keeler to August Vollmer, March 19, 1934, AVP.
84. Leonarde Keeler to August Vollmer, July 29, 1936, AVP.

85. Leonarde Keeler to August Vollmer, February 26, 1937, AVP.
86. Ibid.
87. Leonarde Keeler to August Vollmer, June 21, 1938, AVP.
88. Ibid.
89. Ibid.
90. Leonarde Keeler to August Vollmer, May 31, 1938, AVP.
91. Leonarde Keeler to August Vollmer, August 21, 1937, AVP.
92. Leonarde Keeler to August Vollmer, November 13, 1938, AVP.
93. R. E. Allen "Lie Detector Pays," *The American City* 54, October 1939, 15.
94. "Lie Detectors for Employees," *Business Week,* September 16, 1939, 36–37; Marston, *The Lie Detector Test,* 155.
95. J. P. McEvoy, "The Lie Detector Goes into Business," *Reader's Digest* (American Edition) 38, February 1941.
96. Alva Johnston, "The Magic Lie Detector I," *Saturday Evening Post* 216, April 15, 1944; Leonarde Keeler to August Vollmer, May 2, 1944, AVP.
97. "Hesse Gems Found in Station Locker," *New York Times,* June 9, 1946, 1.
98. Molly Rhodes, "Wonder Woman and her Disciplinary Powers: The Queer Intersection of Scientific Authority and Mass Culture," in *Doing Science and Culture: How Cultural and Interdisciplinary Studies Are Changing the Way We Look at Science and Medicine,* ed. Roddey Reid and Sharon Traweek (New York: Routledge, 2000), 99.
99. Ken Alder, "A Social History of Untruth: Lie Detection and Trust in Twentieth-Century America," *Representations* 80 (2002): 11, 14.
100. Charles Thorpe and Steven Shapin, "Who Was J. Robert Oppenheimer?: Charisma and Complex Organization," *Social Studies of Science* 30 (2000): 580.
101. Max Weber, "The Sociology of Charismatic Authority / The Nature of Charismatic Authority and Its Routinization," in *The Celebrity Culture Reader,* ed. P. David Marshall (London: Routledge, 2006), 56.
102. Ibid.

Conclusion. The Hazards of the Will to Truth

Epigraphs. Nietzsche, *Beyond Good and Evil: Prelude to a Philosophy of the Future,* trans. R. J. Hollingworth (London: Penguin, 1973), 33. Philip K. Dick, *Blade Runner (Do Androids Dream of Electric Sheep)* (New York: Ballantine Books, 1968), 35.

1. "Lie Detector Seals Doom of Murderer," *New York Times,* March 2, 1937, 44.
2. "Lie Detector: Marks in Ink Final Judges for Murder Case," *Newsweek* 9, March 13, 1937, 34.
3. "Polygraph Proof: Illinois Murderer Dies Because of Governor's Belief in Test," *Literary Digest* 123, March 13, 1937, 9–10.
4. "Polygraph Proof."
5. "Lie Detector Seals Doom."
6. "'Lie detector' Gets $10 from 2 boys," *New York Times,* June 8, 1937, 27.
7. Ibid.
8. The *New York Times* reported on a number of "fake" lie detectors during the 1930s. In

one case, a confession was obtained "with [a] contraption of radio parts and hot peppers" ("'Lie Detector' Traps Philadelphia Youth"). In another, a school principal was forced to destroy his contraption—a "black box equipped with dials and electric bulbs"—despite having obtained a confession from a thief of a pair of gloves ("Principal His 'Lie Detector,'" *New York Times*, September 25, 1936, 5).

9. Leonarde Keeler, "Debunking the 'Lie-Detector,'" *Journal of Criminal Law, Criminology and Police Science* 25 (1934–35): 153.

10. Ibid., 158

11. Ibid.

12. Leonarde Keeler, "A Method For Detecting Deception," *American Journal of Police Science* 1 (1930): 48.

13. Such apparently extraneous procedures have since become codified as correct practice in polygraph textbooks. These texts advise polygraphers to pay careful attention to the design of the examination room, the placement and appearance of the machine, and even to their mannerisms and body language. James Allan Matté, *The Art and Science of the Polygraph Technique* (Springfield, IL: Charles C. Thomas, 1980); John E. Reid and Fred E. Inbau, *Truth and Deception: The Polygraph ("Lie Detection") Technique* (Baltimore: Williams and Wilkins, 1977), 5–7.

14. Ken Alder, "A Social History of Untruth: Lie Detection and Trust in Twentieth-Century America," *Representations* 80 (2002).

15. Stephanie A. Shields, "Passionate Men, Emotional Women: Psychology Constructs Gender Difference in the Late 19th Century," *History of Psychology* 10, no. 2 (2007): 92–110.

16. Gaston Bachelard, *The Formation of the Scientific Mind: A Contribution to a Psychoanalysis of Objective Knowledge*. Intro., trans., and annotated by Mary McAllester Jones (Manchester: Clinamen Press, 2002). First published in 1938.

17. Michael Billig, *Ideological Dilemmas: A Social Psychology of Everyday Thinking* (London: Sage, 1988).

18. Oscar Wilde, "The Decay of Lying," in *The Writings of Oscar Wilde*, ed. Isobel Murray (New York: Oxford University Press, 1989), 215–40.

19. Thomas F. Gieryn, "Boundary-Work and the Demarcation of Science from Non-Science: Strains and Interests in Professional Ideologies of Scientists," *American Sociological Review* 48, no. 6 (1983): 781–95.

20. David Garland, "Of Crimes and Criminals: The Development of Criminology in Britain," in Mike Maguire, Rod Morgan, and Robert Reiner, *The Oxford Handbook of Criminology*, 2nd ed. (Oxford: Oxford University Press, 1997), 11–56.

21. Alder, "A Social History of Untruth," 2.

22. Quoted in Ken Alder, *The Lie Detectors: The History of an American Obsession* (New York: Free Press, 2007), 68.

23. Alder, *The Lie Detectors*, 122; Alder, "A Social History of Untruth," 12.

24. Frederick L. Collins, "The Future Looks Dark for Liars," *Collier's*, August 16, 1924, 26.

25. Michael L. Fleisher, *The Encyclopedia of Comic Book Heroes* 2, *Wonder Woman* (New York: Collier Books, 1976), 210.

26. Marston's view of the lie detector must be seen in the light of his assertion that freedom could only be obtained through "submission to loving superiors." He regarded the ma-

chine not as an instrument of bondage-as-torture, but rather a quasi-therapeutic technology that promoted freedom through its ability to produce truth. See Geoffrey C. Bunn, "The Lie Detector, *Wonder Woman* and Liberty: The Life and Work of William Moulton Marston," *History of the Human Sciences* 10 (1997): 91–119.

27. Fleisher, *The Encyclopedia of Comic Book Heroes*, 176.

28. Derek Hook, "Analogues of Power: Reading Psychotherapy through the Sovereignty-Discipline-Government Complex," *Theory and Psychology* 13 (2003): 605–28.

29. Michel Foucault, *Discipline and Punish: The Birth of the Prison*, trans. Alan Sheridan (New York: Vintage Books, 1979).

30. Peter J. Hutchings, "Spectacularizing Crime: Ghostwriting the Law," *Law and Critique* 10 (1999): 27.

31. Jeremy Bentham, *The Panopticon Writings*, ed. Miran Bozovic (London: Verso, 1995), 101.

32. Quoted in Hutchings "Spectacularizing Crime," 42.

33. Ibid., 44.

34. Ibid., 35.

35. Peter J. Hutchings, *The Criminal Spectre in Law, Literature and Aesthetics* (London: Routledge, 2001), 1.

36. John Bender, *Imagining the Penitentiary: Fiction and the Architecture of Mind in Eighteenth-Century England* (Chicago: University of Chicago Press, 1987), 197–98.

37. Philip Smith, "Narrating the Guillotine: Punishment Technology as Myth and Symbol," *Theory, Culture and Society* 20 (2003): 27.

38. Ibid., 43.

39. Guy Debord, *The Society of the Spectacle*, trans. Donald Nicholson-Smith (New York: Zone Books, 1995), chap. 3. First published 1967.

40. Philip K. Dick, *Blade Runner (Do Androids Dream of Electric Sheep)* (New York: Ballantine Books, 1968), 35.

41. *Blade Runner: The Director's Cut*, Warner Home Video, 1993. First released 1982.

42. Dick, *Blade Runner*, 41.

43. Paul M. Sammon, *Future Noir: The Making of Blade Runner* (New York: HarperPrism, 1996), 106–7.

44. Retrieved from http://en.wikipedia.org/wiki/Voight-Kampff_machine (accessed August 31, 2009).

45. Deborah Jermyn, "The Rachel Papers: In Search of *Blade Runner*'s Femme Fatale," in *The Blade Runner Experience: The Legacy of a Science Fiction Classic*, ed. Will Brooker (London: Wallflower Press, 2005), 159.

46. Ibid, 161.

47. Scott Bukatman, *Blade Runner* (London: British Film Institute, 1997), 80–83.

ESSAY ON SOURCES

The immensely useful *Reader's Guide to Periodical Literature: An Author and Subject Index* (New York: H. W. Wilson, 1901–) enabled me to locate a variety of lie detector and polygraph-related articles published in obscure magazines and journals. Useful bibliographies are Norman Ansley and Frank Horvath, *Truth and Science: A Comprehensive Index to International Literature on the Detection of Deception and the Polygraph (Lie Detector) Technique* (Linthicum Heights, MD: American Polygraph Association, 1977); Earleen H. Cook, *The Lie Detector: Its Use in Law and Business* (Monticello, IL: Vance Bibliographies, 1981); and Verna Casey, *Lie Detectors and Detection: A Selected Bibliography, 1985-1987* (Monticello, IL: Vance Bibliographies, 1988).

The papers of Eloise, Charles, and Leonarde Keeler, August Vollmer, and John Larson are in The Bancroft Library, University of California, Berkeley. William Moulton Marston's papers (mainly concerning *Wonder Woman*) are in the Dibner Collection at the Smithsonian Institution, Washington, DC. A comprehensive list of archival sources is in Ken Alder's excellent *The Lie Detectors: The History of an American Obsession* (New York: Free Press, 2007), 275-77. Early histories of the lie detector include Paul V. Trovillo, "A History of Lie Detection," pts. 1 and 2, *Journal of Criminal Law and Criminology* 29 (1939): 848-81; 30 (1939): 104-19; John A. Larson, "The Lie Detector: Its History and Development," *Journal of the Michigan State Medical Society* 37 (1938): 893-97; and Eugene B. Block, *Lie Detectors: Their History and Use* (New York: David McKay Co., 1977).

On Leonarde Keeler see Eloise Keeler, *The Lie Detector Man: The Career and Cases of Leonarde Keeler* (Boston: Telshare Publishing, 1984). On William Moulton Marston and Wonder Woman, see Geoffrey C. Bunn, "The Lie Detector, Wonder Woman and Liberty: The Life and Work of William Moulton Marston," *History of the Human Sciences* 10 (1997): 91-119; Les Daniels, *Wonder Woman: The Complete History* (London: Titan Books, 2000); Gerard Jones, *Men of Tomorrow: Geeks, Gangsters and the Birth of the Comic Book* (London: William Heinemann, 2005). For a selection of Marston's stories see Charles Moulton [William Moulton Marston] *Wonder Woman* (New York: Holt, Rinehart and Winston and Warner Books, 1972, first published 1943-49); William Moulton Marston and H. G. Peter, *Wonder Woman Archives*, vols.1-6 (New York: DC Comics, 1998-2010, first published 1941-45).

Important primary source lie detector material for students of history or science studies include John A. Larson, *Lying and its Detection: A Study of Deception and Deception Tests*

(Glen Ridge: Patterson Smith, 1932); William Moulton Marston, *The Lie Detector Test* (New York: Richard R. Smith, 1938); Paul L. Wilhelm and F. Donald Burns, *Lie Detection with Electrodermal Response*, 5th ed. (Michigan City, IN: B and W Associates, 1954); John E. Reid and Fred E. Inbau, *Truth and Deception: The Polygraph ("Lie Detection") Technique* (Baltimore: Williams and Wilkins, 1977); Robert J. Ferguson Jr., *The Scientific Informer* (Springfield, IL: Charles C. Thomas, 1971); Chris Gugas, *The Silent Witness: A Polygraphist's Casebook* (Englewood Cliffs, NJ: Prentice-Hall, 1979); James Allan Matté, *The Art and Science of the Polygraph Technique* (Springfield, IL: Charles C. Thomas, 1980); Murray Kleiner, *Handbook of Polygraph Testing* (San Diego: Academic Press, 2002); John F. Sullivan, *Gatekeeper: Memoirs of a CIA Polygraph Examiner* (Dulles, VA: Potomac Books, 2008).

More recent studies on the polygraph from a scientific standpoint include Anthony Gale, ed., *The Polygraph Test: Lies, Truth and Science* (London: Sage, 1988); Gershon Ben-Shakhar and John J. Furedy, *Theories and Applications in the Detection of Deception: A Psychophysiological and International Perspective* (New York: Springer-Verlag, 1990); David T. Lykken, *A Tremor in the Blood: Uses and Abuses of the Lie Detector*, 2nd ed. (New York: Basic Books, 1998); National Research Council, *The Polygraph and Lie Detection: Committee to Review the Scientific Evidence on the Polygraph* (Washington, DC: National Academies Press, 2003); Aldert Vrij, *Detecting Lies and Deceit: Pitfalls and Opportunities*, 2nd ed. (Chichester: Wiley, 2008).

Essential criminology and criminal anthropology primary sources include Enrico Ferri, *Criminal Sociology*, trans. W. D. Morrison (London: T. Fisher Unwin, 1895); Arthur Macdonald, *Criminology* (New York: Funk and Wagnalls, 1893); Cesare Lombroso and Guglielmo Ferrero, *The Female Offender* (New York: D. Appleton, 1895); Cesare Lombroso, *Criminal Man According to the Classification of Cesare Lombroso, Briefly Summarized by His Daughter Gina Lombroso Ferrero, with an Introduction by Cesare Lombroso* (New York: G. P. Putnam's Sons, 1911); Cesare Lombroso, *Criminal Man*, trans. with a new intro. by Mary Gibson and Nicole Hahn Rafter (Durham, NC: Duke University Press, 2006); Hugo Münsterberg, *On the Witness Stand: Essays on Psychology and Crime* (New York: Clark Boardman, 1927, first published 1908); Havelock Ellis, *The Criminal*, 5th ed. (London: Walter Scott Publishing Co., 1914); Hans Gross, *Criminal Psychology: A Manual for Judges, Practitioners, and Students*, trans. Horace M. Kallen (Boston: Little, Brown and Co., 1911, first published 1905); Charles B. Goring, *The English Convict: A Statistical Study* (London: H.M.S.O., 1913; reprint, Montclair, NJ: Patterson Smith, 1972).

The history of criminology and criminal anthropology has experienced a scholarly renaissance since the appearance of Michel Foucault's paradigm shifting *Discipline and Punish: The Birth of the Prison*, trans. Alan Sheridan (New York: Vintage Books, 1979). Stephen Jay Gould's *The Mismeasure of Man*, rev. and enl. ed. (New York: W. W. Norton, 1996) is a useful place to start reading about Lombroso's criminal anthropology. Important sources also include Arthur E. Fink, *Causes of Crime: Biological Theories in the United States* (Philadelphia: University of Pennsylvania Press, 1938); Robert Nye, "Heredity or Milieu: The Foundations of European Criminological Theory," *Isis* 67 (1976): 335–55; Martin J. Wiener, *Reconstructing the Criminal: Culture, Law, and Policy in England, 1830–1914* (Cambridge: Cambridge University Press, 1990); Lucia Zedner, *Women, Crime and Custody in Victorian England* (Oxford: Clarendon Press, 1991); Marie-Christine Leps, *Apprehending the Criminal: The Production of De-*

viance in Nineteenth-Century Discourse (Durham, NC: Duke University Press, 1992); Alison Young, *Imagining Crime: Textual Outlaws and Criminal Conversations* (London: Sage, 1996); Nicole Hahn Rafter, *Creating Born Criminals* (Champaign: University of Illinois Press, 1997); Richard F. Wetzell, *Inventing the Criminal: A History of German Criminology, 1880–1945* (Chapel Hill: University of North Carolina Press, 2000); Peter J. Hutchings, *The Criminal Spectre in Law, Literature and Aesthetics* (London: Routledge, 2001); Mary Gibson, *Born to Crime: Cesare Lombroso and the Origins of Biological Criminology* (Westport, CT: Praeger, 2002); David G. Horn, *The Criminal Body: Lombroso and the Anatomy of Deviance* (London: Routledge, 2003); Peter Becker and Richard F. Wetzell, *Criminals and their Scientists: The History of Criminology in International Perspective* (Cambridge: Cambridge University Press, 2006); Neil Davie, *Tracing the Criminal: The Rise of Scientific Criminology in Britain, 1860–1918* (Oxford: Bardwell Press, 2006); David Garland, "Of Crimes and Criminals: The Development of Criminology in Britain," in Mike Maguire, Rod Morgan, and Robert Reiner, *The Oxford Handbook of Criminology*, 2nd ed. (Oxford: Oxford University Press, 2007).

On degeneration see Daniel Pick, *Faces of Degeneration: A European Disorder, c 1848–1918* (Cambridge: Cambridge University Press, 1989); Kelly Hurley, *The Gothic Body: Sexuality, Materialism, and Degeneration at the Fin-de-Siècle* (Cambridge: Cambridge University Press, 1996). For phrenology see David de Giustino, *Conquest of Mind: Phrenology and Victorian Social Thought* (London: Croom Helm, 1975); Roger Cooter, *The Cultural Meaning of Popular Science: Phrenology and the Organization of Consent in Nineteenth-Century Britain* (Cambridge: Cambridge University Press, 1984); John Van Wyhe, *Phrenology and the Origins of Victorian Scientific Naturalism* (Aldershot: Ashgate, 2004); and Nicole Hahn Rafter, "The Murderous Dutch Fiddler: Criminology, History and the Problem of Phrenology," *Theoretical Criminology* 9, no. 1 (2005): 65–96.

The most recent scholarship on the history of the lie detector includes Ken Alder, "To Tell the Truth: The Polygraph Exam and the Marketing of American Expertise," *Historical Reflections* 24 (1998): 487–525; Margaret Gibson, "The Truth Machine: Polygraphs, Popular Culture and the Confessing Body," *Social Semiotics* 11, no. 1 (2001): 61–73; Ken Alder, "A Social History of Untruth: Lie Detection and Trust in Twentieth-Century America," *Representations* 80 (2002): 1–33; Ken Alder, *The Lie Detectors: The History of an American Obsession* (New York: Free Press, 2007); Geoffrey C. Bunn, "Spectacular Science: The Lie Detector's Ambivalent Powers," *History of Psychology* 10, no. 2 (2007): 156–78.

INDEX

Adams, Frederick, 162
Adler, Herman M., 118, 165
Albrecht, Adalbert, 44, 46, 63, 64
Alder, Ken, 183
American Polygraph Association, 188
Ames, Aldrich, 5
Appel, Charles A., 169
Aristotle, 52
Aschaffenburg, Gustav, 27
atavism, 17, 22, 27, 39, 48, 63
Austin, Alfred, 75, 77
automatograph, 100, 108
Ayur Veda, 134

Backster, Cleve, 4–5
Baer, Abraham, 27
Balmer, Edwin, 110–12, 117, 142, 180
Barthes, Roland, 134, 152
Beard, George Miller, 54
Beccaria, Cesare, 8
Bell, Alexander Graham, 126
Benedikt, Moritz, 38
Bentham, Jeremy, 8, 186, 187
Berardi, Vito Antonio, 61
Berkeley Psychograph, 171
Bernaldo de Quirós y Pérez, Constancio, 26
Bernard, Claude, 67, 71
Bertillon, Alphonse, 25
Bertino, Michel, 67
Blackwood, Algernon, 110
Blade Runner, 189–92
Blazenzits, Joe, 166
Bleuler, Eugen, 27, 49, 99

Block, Eugene, 120
blushing, 42, 65–66, 190
Boies, Henry, 41–42, 45
Booth, John Wilkes, 169
Bordoni, Ernesta, 43
Boring, Edwin Garrigues, 98
born criminal: abandoned by criminology, 76, 95, 133, 180; in criminology, 20, 22, 23, 29, 46, 177; in England, 35, 36; Kraepelin, 27; lack of emotion of, 72; Lombroso's notion of, 18, 22, 28, 48–49, 177; rejected by detective fiction, 112, 180; skepticism concerning, 25, 27, 28, 83, 108; in the United States, 38, 105; women, 43, 59, 62, 63, 65, 73
Bourget, Paul, 78
Bray, Charles, 16
Bridges, Frederick, 13–15
Brinton, Daniel Garrison, 39–40, 59
Broca, Pierre Paul, 21
Burgess, Thomas Henry, 65
Burke, William, 12
Burtt, Harold E., 139
Butler, Josephine, 61

Call Northside 777, 172
Canary Murder case, 166
Capillary electrometer, 125
Capone, Al, 166
cardiograph, 67, 71, 110, 115
Cardio-Pneumo-Psychogram, 122, 125
Casey, Patrick, 175
Chaplin, Charlie, 164
Charcot, Jean-Martin, 96

charismatic authority: of criminal man, 187; of criminology, 31, 178, 189; defined, 48–49, 134, 172–73; of lie detector, 31, 183, 192; of lie detector experts, 143, 172–73, 181–82; of Lombroso, 31, 48–50, 178; and sovereign power, 186–87
Chesterton, Gilbert Keith, 94
chronoscope, 68, 98, 102, 103, 108, 110, 111, 114, 118, 121, 122
Clouston, Thomas Smith, 23–24, 33–34, 55
Colajanni, Napoleone, 25
Collins, Frederick, 118
Collins, Wilkie, 75, 76, 77
Combe, George, 13, 15
Comte, Auguste, 21, 54
Conan Doyle, Arthur, 80, 179
Coogan, Jackie, 146
Corday, Charlotte, 42
craniology, 16–17, 21
crime, as normal feature of society, 179
criminal: as a biological entity, 17–21, 23–24, 29, 33, 178; female, 73, 178, 179; as the focus of criminology, 9–11, 20, 23, 28, 35, 178; insensitive to pain, 64
criminal anthropology: absence in Great Britain, 35, 37, 47; in Austria, 28; as contradictory discourse, 49; critique of, 75–93, 108, 124; fetish for instrumentation, 69, 72; in France, 24–25; in Germany, 27–28, 47; influence in literature, 77–81, 108, 115; interest in female offenders, 60, 61, 73, 91, 148; and Lombroso, 46, 49, 50; opponents of, 25, 32, 38; origins and emergence of, 18, 41, 56, 179; and penal codes, 30; and politics, 30; rhetorical modes of, 43–45, 178; search for stigmata, 4, 23; sensationalism of, 42, 44; and social policy, 24, 40; in Spain, 25–26, 47; in the United States, 38–40, 88–91, 105–6
criminal insanity, 20
criminal jurisprudence, 15, 20
Criminal Man (Lombroso), 20, 21, 22, 27, 31, 41, 47, 177, 187
criminology: abandons the born criminal, 76, 108, 110, 180; in Britain, 33, 34, 35, 37; development of, 9–10, 20, 23, 28, 30–31, 52, 68, 73; dilemmas of, 189; as dilemmatic discourse, 178; governmental project of, 182; Lombrosian project of, 182
cruentation, 8, 196n5

Dallemagne, Jules, 47
Darrow Photopolygraph, 171

Darwin, Charles, 21, 54, 66, 67, 79, 82
Darwin, Leonard, 37
de Aramburu, Félix, 25
Deception Tests Service Company (Berkeley), 171
degeneration theory, 20, 21–22, 27, 35, 40, 60, 78–80, 84, 105
de Mille, Agnes, 164
de Mille, Cecil B., 164
Dent, Max, 174
detective fiction, 77, 81, 110, 180, 181
Dick, Philip K., 174, 189
Dickens, Charles, 76, 77
Dick Tracy, 172
Distant, William Lucas, 58
Dorado Montero, Pedro, 26
Dracula (Stoker), 39, 80
Duchenne de Boulogne, Guillaume-Benjamin-Amand, 67, 96
Dugdale, Richard, 79, 87
Dunlap, A., 129
Dyche, William, A., 168
dynamometer, 108

Edison, Thomas, 126, 127
electric psychometer, 94, 95, 103, 104, 106, 107, 111, 112
Ellis, Henry Havelock, 35, 36, 38, 44, 45, 59, 63, 66
Elmira State Reformatory, New York, 39, 45, 86
emotiograph, 118, 125
Erasistratus, 134
ergograph, 108
eugenics, 35, 107

Falco, Giovanni, 43
Falret, Jules, 58
feeblemindedness, 107, 108, 147
female body, as focus of criminology, 178
Female Offender, 59–60
female offenders, 65
Féré, Charles, 25, 66, 96, 97
Ferrer, Francisco, 26
Ferrero, Guglielmo, 42–43, 59, 61, 62, 63, 64, 65
Ferri, Enrico, 22, 38, 49, 71, 88
Feuerbach, Paul Johann Anselm von, 8
Feyerabend, Paul, 1
Fibbograph, 100
Fink, Arthur E., 108
Fletcher, Ronald, 38–39

Földes, Béla, 59, 61
Ford, Henry, 126, 127
Fordham psycho-galvanometer, 138
Foucault, Michel, 186, 187
Fox, Long, 36
Freeman, Austin Richard, 110
Freud, Sigmund, 51, 55, 96
Futrelle, Jacques Heath, 114

Gage, Phineas, 12
Gall, Franz Josef, 12–13, 21
Galton, Francis, 35, 37, 64
galvanometer, 96, 98, 99, 100, 105, 110, 111, 112, 114, 127, 133, 144, 182
Garofalo, Raffaele, 22–23, 32, 38, 64
Gaudenzi, Carlo, 18
Gautier, Alfred, 49
Geddes, Patrick, 54–55
Gesell, Arnold, 132
Gilman, Charlotte Perkins, 53–54
Goddard, Calvin, 150, 166, 169, 170
Goddard, Henry Herbert, 106–8
Golden Lasso of Truth (Wonder Woman), 158, 184, 185, 188
goniometer, 68–69
Goring, Charles, 25, 37, 83, 108, 109
Gothic novel, 77–81, 83, 109, 110, 179
Grable, Betty, 146
Griffiths, Arthur, 37, 47
Griffiths, G. B., 37
Gross, Hans, 28, 219n120
guillotine, 187

Haeckel, Ernst, 21
Hare, William, 12
Harwood, Jack, 1
Hauptmann, Bruno, 146, 166, 227n79
Healey, William, 108
Hegel, Georg Wilhelm Friedrich, 53
Heinrich, Edward Oscar, 124
Henderson, Charles, 40
Hesse-Darmstadt jewels, 172
Hill, Anita, 5
Hodgson, William Hope, 110
Homo criminalis. See *Criminal Man* (Lombroso)
Hooton, Ernest A., 45
Hoover, J. Edgar, 230n22
Horner, Henry, 174
Howard, John, 8

Huxley, Thomas Henry, 82
hysteria, 56–58, 80, 97, 179

ideological dilemmas, 46, 82, 178, 181–84, 186, 188–89
Inbau, Fred, 128, 134, 139, 151, 167, 184
inspiration-expiration ratio, 115
Institute for Juvenile Research (Chicago), 165
intelligence tests, 107

"Jack the Ripper" murders, 77
Jastrow, Joseph, 38, 70
Johnston, Alva, 150
Johnstone, Edward R., 106, 107
Jung, Carl, 96, 97, 98, 99, 131, 158
Juska, Edward, 138

Kant, Immanuel, 55
Keeler, Charles Augustus, 161–62, 164, 165
Keeler, Eloise, 120, 126, 136, 154, 166–69
Keeler, Leonarde: in *Call Northside 777*, 172; canary murder case, 166; charismatic authority of, 172; correspondence with August Vollmer, 128–29, 140, 164, 169–72; fame of, 154, 170–72; and Hollywood, 164, 172; as inventor of lie detector, 117–18, 120, 124, 125, 131, 133, 138, 144, 154, 168, 175–76, 182; and lie detector business, 171–72, 182, 184; lie detector technique of, 142, 143, 144, 154; on Marston, 128–29, 131; patent on lie detector, 133, 165, 223n79, 223n83; in photographs of lie detector tests, 136, 142, 144, 150, 165, 168; public recognition for work of, 167; and Rappaport case, 142–43, 144, 147, 174–77; relationship with Fred Inbau, 130; relationship with John Larson, 128, 130, 140, 184; relationship with August Vollmer, 164 (*see also* correspondence with August Vollmer); youth of, 162, 164
Keeler Polygraph, 149; in magazine articles, 135, 138, 140, 145, 154, 166, 168, 171, 172; synonymous with lie detector, 128, 137, 149, 165, 167, 183
Kellor, Frances, 87–93, 180, 214n91
kimegraph, 121
Kipling, Rudyard, 78
kissing, 44
Klein, Charles, 225n13
Koch, Julius, 27
Kraepelin, Emil, 27
Kurella, Hans, 27, 47
kymograph, 91, 115, 127, 180

Index

Lacassagne, Alexandre, 24, 25
Lapicque, Louis, 105
Larson, John Augustus, 116–18, 120, 122–25, 128–33, 139, 140, 148, 150, 183–84, 186; relationship with Leonarde Keeler, 128, 130, 140, 184; relationship with August Vollmer, 164
Lavine, Emanuel H., 135
Lee, C. D., 130
Leibowitz, Samuel, 148
Leonarde Keeler, Inc., 171
Leonarde Keeler Polygraph Institute, 5
Lewis, A. A., 143, 146
lie detector: accuracy statistics of, 138, 139; aim to produce confessions, 177; as alternative legal system, 147, 180, 183, 191, 218n109; animosity between pioneers, 128–31; as black box, 124, 140, 142–43, 183; conditions of possibility for, 192; as contradictory discourse, 5, 151, 177, 181–85, 188; discursive architecture of, 183; as dream of criminology, 148; emergence of, 179, 180–82; as erotic scene, 152, 186; fails with certain subjects, 147; first use of term, 95, 108, 124, 182; as function of privileging of the lie, 124–25; as gendered practice, 2, 151; images of, 150; interest of popular press in, 137, 151–52; intimidating reputation of, 141, 148, 183; invention of, 115, 117, 120, 124, 172; as invention, 124, 126, 131, 133, 135, 137, 176, 182, 224n113; inventor of, 116–33, 173, 182; in law court, 155, 229n5; "male gaze" of, 148–50; mystique surrounding, 127, 137, 140, 141, 173, 188; myth of invention of, 173, 182, 224n113; naming of, 125; as opposite of "third degree," 164, 168, 183; as popular culture construction, 172, 180; as possessing agency, 145; as possessing magical properties, 143–44, 172, 183; price of, 171; as public relations tool, 168; to solve romantic problems, 160, 183, 184, 186; as tool of medical diagnosis, 184, 186; volunteering to take test, 146; women as ideal subjects of, 151
lietector, 125
Lindbergh baby kidnapping, 5, 146, 159, 166
Lombroso, Cesare: ambitions for criminology, 20–21; charismatic authority of, 46–50, 178; discusses politics, 26, 32; as father of criminology, 20, 31, 46, 134; and female offenders (see *Female Offender*); on genius, 43, 82; Homo criminalis (*L'uomo delinquent*), 21, 22, 27, 47; on hysterica, 56; influence of, 34–38, 41, 46, 47, 78, 80–83, 101, 105; influences on, 21, 22, 41, 68, 78, 178; and instrumentation, 18, 68, 70, 72–73; rhetorical modes of, 42–46, 48, 178; role in creating criminal anthropology, 18, 30–38, 41, 47, 178; theory criticized, 25, 28, 37, 38, 45, 83–85, 88–89, 110; theory of the born criminal, 18, 21, 22, 23, 49, 64, 66, 70, 177
Lombroso-Ferrero, Gina, 44
Londes, Nick, 146
Lopez, Marquita, 156
Ludwig, Carl Friedrich Wilhelm, 67
L'uomo delinquent. See *Criminal Man* (Lombroso)
Lydson, G. Frank, 40
lying, 75, 181
Lykken, David, 2

MacDonald, Arthur, 40, 69, 73, 84–85, 88, 105
MacHarg, William, 110, 112, 117, 142, 180
Mackenzie, George S., 16
Mantegazza, Paolo, 61, 88
Marat, Jean-Paul, 42
Marconi, Guglielmo, 127
Marey, Étienne-Jules, 67, 71
Marston, William Moulton: charisma of, 154, 155, 159, 182–83; creates *Wonder Woman*, 158, 172, 185; criticized notion of invention, 126, 133; dominance and submission theory of, 156, 158, 185; FBI file on, 230n22; as inventor of lie detector, 117–21, 131, 132, 155, 158, 172, 182; on Keeler, 128; lie detector technique of, 122–24, 139, 147, 150, 158; *The Lie Detector Test* (1938), 118, 122, 128, 129, 135–36, 159, 184; managed lie detector's contradictions, 184–86; offers Hauptmann a lie detector test, 146; as popular psychologist, 156–58, 184; against "third degree," 137; use of lie detector extended beyond crime, 154–55, 160–61, 184–86; uses lie detector in court of law, 120, 155; on women as ideal subjects, 151; work criticized by others, 130, 139, 184
Marston Systolic Blood Pressure Deception Test, 120–23, 126, 139, 149, 155–57, 159
Marzolo, Paolo, 21, 41
Massee, Burt, 165, 170
Matté, James Allen, 151
Maudsley, Henry, 17, 19, 55, 57, 59, 76
Mayhew, Henry, 10, 18, 24, 33, 65–66

McCarty, Dwight G., 120
McKim, W. Duncan, 40
McLaughlin, John, 177
McLean Hamilton, Allan, 95, 96, 104
Mead, Syd, 190
Mendaxophone, 100
mistrust, between lie detector pioneers, 131, 184
Mobile Crime Detection Laboratory, 167
Moffett, Cleveland, 115, 180
Montaigne, Michel de, 53
Montero, Dorado, 46
moral insanity, 11, 41
Morel, Bénédict Augustin, 10–11, 21
moron, 108, 147
Morrison, William D., 36–37, 83–84
Mosso, Angelo, 67, 73, 132, 134
Müller, E. K., 97
Münsterberg, Hugo, 99–104, 111, 113, 115, 120, 126, 131, 132, 137, 157–58, 225n13
Musolino, Guiseppe, 43

Näcke, Paul, 27
National Committee for Mental Hygiene, 155
Niceforo, Alfredo, 58
Nietzsche, Friedrich, 51, 53, 56, 174, 189
Nordau, Max, 46

Olson, Walter, 165
Orchard, Harry (Albert Horsley), 102, 103, 131
Ottolenghi, Salvatore, 20–21, 43, 59, 60, 65
Owen, M. E., 66

Panizza, Bartolomeo, 21
Panopticon, 187
Parsons, Philip A., 40
Partridge, George Everett, 66
Pearson, Karl, 25
penal policy, 8, 20
penology, 20, 33, 40
Peterson, Frederick, 97–98, 99, 103
phrenology, 10–16, 20, 21, 35, 40
Pickford, Mary, 164
Pie-crustograph, 100
Pitrè, Giuseppe, 59
Plato, 52
plethysmograph, 72–73, 92, 98, 100, 108, 110, 111, 113, 114, 118, 142

Pneumo-Cardio-Sphygmogalvanograph, 125
pneumo-cardio-sphygmometer, 117
pneumograph, 98, 100, 108, 110, 111, 113, 118, 127, 133, 142, 156, 180, 182
Pollock, Dorothy, 170
polygraph. *See* lie detector
Price, Rosemary, 125
Prichard, James, 11
Pringle, Henry, 140
prostitute: criminology's obsession with, 24, 43, 57, 61–63, 179; as embodiment of criminality, 60–61, 179
prostitution, 10, 61–63, 88
psychoanalysis, 83
psychogalvanometer, 125, 140, 144
psychometer. *See* electric psychometer
psychopathic subjects, 147
pulp fiction, 109–15

Quetelet, Adolphe, 9–10, 61

Rappaport, Joseph, 142–43, 144, 147, 174–77
Ray, Isaac, 57
Reeve, Arthur Benjamin, 95, 112, 113, 115, 117, 180
Reid, John E., 151
Rittenberg, Max, 110
Rizzo, Ernie, 1–2
Road Hill House murder, 76
Robinson, Henry Morton, 135
Rosny aîné, J.-H. (Joseph-Henri Honoré Boex), 78
Rousseau, Jean-Jacques, 55
Rush, Benjamin, 8

Salillas, Rafael, 26, 47
schizophrenia, 99
Schopenhauer, Arthur, 55
Scientific Crime Detection Laboratory (Northwestern University, Chicago), 139, 140, 154, 155, 165, 167, 168, 169, 171, 175; finances of, 170
Scott, Orlando F., 125, 151
Scripture, E. W., 98
sensation novel, 77
Severy, Melvin L., 115, 180
sham experiment, 104
Sighele, Scipio, 59
Simpson, O. J., 1–2, 5
Simpsons, The, 191

Skey, Frederic Carpenter, 58
Sloan, Charles, 118
Smith, Joshua Toulmin, 16
Smith, Philip, 187
Sommer, Robert, 27, 97
soul machine, 94, 103, 104, 111, 124, 147, 180
spectacle, 50, 188–89
Spencer, Herbert, 21, 54
sphygmograph, 67, 70, 71, 100, 114, 115
sphygmomanometer, 71, 118, 121, 122, 127, 133, 156, 182
sphygmophone, 115
Steunenberg, Frank, 102
Stevenson, Robert Louis, 78, 82, 179
Stewart, Jimmy, 172
Sticker, G., 97
Stoker, Bram, 179
Stone, Thomas, 12
St. Valentine's Day Massacre, 165
Sullivan, T. P., 167
Summers, Walter G., 125, 139, 144
systolic blood pressure deception test (Marston), 120–23, 126, 139, 149, 155–57

Taine, Hippolyte, 78
Taladriz, Alvarez, 26
Tarchanoff, Jean de (Ivan), 97
Tarde, Gabriel, 24, 25, 33, 72, 85
Tarde, Gustave, 45, 47, 49–50
Tarnowsky, Pauline, 62, 88
tattoos, 20, 31, 41, 45, 64, 70, 72, 178
third degree, 101–2, 104, 135, 137, 141, 148, 151, 163, 164, 169, 181, 183, 188, 225n13
Thomas, Clarence, 5
Thomson, J. Bruce, 7, 17–19, 24
Thomson, John Arthur, 54–55
Tobin, William, 166
Topinard, Paul, 24
Trovillo, Paul, 134, 139, 149, 150, 151
Turati, Filippo, 25

Veraguth, O., 97
Vigouroux, Romain, 96
Villella, Giuseppe, 18, 41
Vineland Training School, New Jersey, 107, 108
Voight-Kampff Empathy Test, 189–91
Vollmer, August, 117–18, 120–21, 128–30, 135, 140, 162, 165; and Berkeley Police Department, 118, 120, 162; and "college cops," 121, 138, 164; correspondence with Keeler, 128–29, 140, 164, 169, 170–72; correspondence with Larson, 130, 140; as "father of modern police science," 163; as friend of Keeler family, 162; and Leonarde Keeler, 162; professionalizing mission of, 135, 162–64; role in lie detector's creation, 117–18, 120, 121, 164; against "third degree," 135

Walk, Charles, 114, 116, 124, 180
Waller, Augustus Desiré, 96
Weber, Max, 48, 134, 172–73
Wells, Herbert George, 82, 179
White, Frank M., 103
Wickersham Commission, 135
Wigmore, John Henry, 165, 166
Wilde, Oscar, 78
Williams, H. S., 85–87
Wilson, G., 16
Wilson, Orlando Winfield, 138
Winkler, Gus, 166
Wonder Woman, 158, 172, 184, 185
Woodworth, Robert Sessions, 103
word association test, 97, 103, 111, 113, 114, 121, 122, 123, 124, 132
World's Fair, Chicago, 169, 232n72
Wright, Orville, 126, 127
Wright, Wilbur, 126, 127

Zanardelli Criminal Code (Italy, 1889), 32
Zimmerman, Gustav, 198n59
Zola, Emile, 78